THE MOTION OF MERCURY'S PERIHELION:

A Reevaluation of the Problem and Its Implications for Cosmology and Cosmogony

By

Harold S. Slusher, D.Sc., Ph.D.

and

Francisco Ramirez, M.S.

ICR Technical Monograph No. 11

Institute for Creation Research
El Cajon, California

THE MOTION OF MERCURY'S PERIHELION:
A Reevaluation of the Problem and Its
Implications for Cosmology and
Cosmogony

Copyright © 1984
Institute for Creation Research

Published by:
Institute for Creation Reseach
2100 Greenfield Dr.
El Cajon, California 92021

Library of Congress No. 83-080180
ISBN 0-932766-17-X

Cataloging in Publication Data
Slusher, Harold Schultz
    The motion of Mercury's perihelion:
a reevaluation of the problem and its
implications for cosmology and cosmogony / by Harold S. Slusher and
Francisco Ramirez.
    (ICR monograph, no. 11)

    1.   Mercury (Planet)   I. Ramirez,
Francisco, jt. auth. II. Title.
III. Series.
                    523.41

Printed in the United States of
America.

TABLE OF CONTENTS

FOREWORD

The theory of relativity, as developed by Albert Einstein and his followers, is an attempt to explain the apparent results of certain intricate optical experiments by a complete reconstruction of our fundamental ideas in regard to space and time. This reconstruction, so radically different from the classical fundamental ideas, has come to permeate the disciplines of cosmology and cosmogony as astronomers try to understand the universe.

From the beginning of thought about the matter, time and space have been considered as independent: time flowing on uniformly regardless of innumerable bodies in the universe and of their motions here and there through space. An interval of time has always been regarded as the same under all conditions and throughout all space. With Einstein all this changed; space and time were bound together; space could not exist without time, and time changes with space and with the motions of the material therein. According to the theory of relativity the interval of time varies from place to place; it is different for the person at rest and for the pilot; it would appear longer for an astronomer on Venus, and shorter to an observer on the slow moving Uranus.

Motion is the ordinary basis of our conception of space. From our every day experience, confirmed by an endless series of measurements, is derived our basic concept of space: "a general receptacle in which things have their existence." The general container, space, contains everything; it is itself devoid of all material attributes. Each point of space is like each and every other point; each point has one attribute and one attribute only, that of position. There is an essential difference between space and time. Space is reversible but time is not. To sum up the fundamental classical concepts of space and time: space is three-dimensional and reversible; time is one-dimensional and irreversible. Both time and space exist independently, and independent of any material thing, or body.

Everyone concedes that the way things appear in our universe depends upon the point of view of the observer. The apparent brightness of a star, the apparent height of a tree, the apparent transverse motion of a car, all depend on the position. In that sense, all of our knowledge is relative, so that the idea of relativity is nothing strange; it is an old concept in the world of physics and mathematics. However, Einstein did not introduce the concept of relativity in this sense, for his idea of relativity has to do with the very

concepts of time and space
themselves. For example, to compre-
hend the difference between Einstein's
concept of relativity and the tradi-
tional concept, we merely note that
all fixed observers measuring the
height of a tree accept the fact that
the tree possesses a real height which
is the same for all observers regardless of the visual angle that the tree
subtends at the eye of each observer.
However, according to Einstein's concept of relativity it is impossible to
speak of the real dimensions of an
object since these dimensions (and in
fact all measurements of space and
time) depend on how an observer is
moving with respect to the object.

According to Isaac Newton's picture of things, events in the universe
unfold in a uniquely ordered way which
is the same for all observers no matter where they are, or how they may be
moving, so that space and time are
absolute. This means that if any two
observers in our universe arranged
charts showing the same series of
events in the order of their occur-
rence as they saw them, they would
both come up with identical charts.
The special theory of relativity de-
nies this; and it is in this sense
that we must understand the word
"relativity" as it is used by Einstein
for the special theory introduces the
idea that both space and time are
relative concepts and that the

ordering of events in our universe is not unique. If Einstein's relativity is accepted no absolute statements about distances or the time intervals between events can be made. The theory of relativity was developed in two parts: (1) the special theory of relativity, which deals with the laws of physics as they are formulated by observers moving with uniform velocity with respect to each other, and (2) the general theory of relativity which deals with observers in accelerated frames of reference or in gravitational fields.

The following table (after Charles L. Poor) exhibits a few of the principal differences between the so-called classical, or standard theories and the relativity theory, as enunciated by Einstein and his disciples:

	Standard Theories	Einstein Theory
Space:	Independent and absolute.	Dependent: connected to and involved with time.
Time:	Independent and absolute: same everywhere and under all conditions.	Dependent upon space: varies with positions and motions of bodies.
Time intervals:	Identical everywhere and under all conditions.	Vary with the positions and motions of bodies.
Rigid bodies:	Of same dimensions and shape under all conditions of motion.	Vary in size and shape with motion.
Geometry:	Laws and formulas the same everywhere and under all conditions.	Laws and formulas vary under gravitational action of material bodies.
Speed of light:	Constant in space.	Appears the same to every observer whatever his motion.
Ray of light:	Travels in a straight line.	Travels in curved paths under attraction of material bodies.
Gravitation:	Independent of motion of bodies. Due to the attraction of material bodies.	Varies with the speed of bodies. Material bodies warp space, and this "warp" causes motion in bodies.

The relativity theory strikes directly at our fundamental concepts as to the structure of the universe; its conclusions are startling and completely upsetting to our ordinary common sense way of looking at physical and astronomical phenomena. To have such a theory accepted, it would seem that the experiments, cited by its supporters, must be clear-cut and admit of no other solution. The burden of proof should be on the relativist, and it should be clearly shown in each case or experiment that the relativity theory is the necessary and sufficient explanation; it should be established beyond all reasonable doubt, not only that the phenomena can be explained by the relativity theory, but that no other hypothesis or theory can equally well account for the observed facts.

Has this been done? Do the experiments and phenomena cited by Einstein and his followers clearly establish the truth of his theories by excluding, as possible explanations, all other hypotheses and theories? The motion of the perihelion of Mercury was the first proof announced by Einstein and the one most widely quoted.

This monograph is the result of study of the motion of the perihelion of Mercury using Newton's law of gravitation, with no modification, making use of the measurements of the oblateness of the sun as obtained by Charles L. Poor and other investigators. This work indicates that the motion of Mercury's perihelion can be explained very accurately on the basis of the sun's oblateness. An explanation of the motion of Mercury's perihelion on the basis of relativity considering the sun's oblateness would fail since this would necessitate the introduction of another term into the equations from general relativity and, thus, give an incorrect answer.

The theoretical basis of modern cosmological theories is the general theory of relativity and in particular Einstein's field equations. The implications of the results of this monograph are far-reaching since they deny the validity of one of the basic evidences cited for the theory of general relativity and consequently, then, the validity of much of modern cosmology and cosmogony.

I.

INTRODUCTION

After Newton's discovery of the Law of Gravitation, theoreticians have attempted to formulate the past, present, and future motions of the heavenly bodies. The motions that can be predicted from these formulations are the result of long years of hard work and dedication by Johannes Kepler. From information gathered by his mentor Tycho Brahe, Kepler deduced three compact laws of planetary motion. The first of these laws states that the motion of a planet around the Sun can be described by an ellipse. The second law states that a planet sweeps out equal areas in equal increments of time throughout its motion. The last law relates the orbital period of the planet to its mean distance from the Sun.

Kepler's laws, however, are valid only for the ideal case of mutual attraction between two spherically homogeneous bodies. The motions of the planets in the solar system, for example, deviate considerably from the predictions of Kepler's theory. These deviations are due to mutual gravitational interactions between multiple bodies in the solar system and the non-spherical conformation of the Sun. Hence, Kepler's three laws are useful only as a first approximation to the problem of orbital motion within the solar system.

The theory presented here examines the perturbation effects caused by the oblateness of the Sun and the multi-gravitational attractions of the remaining planets on the one taken into consideration. This approach closely follows that performed by Moulton[1] and McCuskey[2], which analyzes the problem of planetary perturbations by taking the special case of two planets and a spherically symmetric Sun (the Classical Perturbation Theory). Therefore, the main difference between their analysis and that of this paper are the considerations of an oblate Sun and the inclusion of multiple gravitational attractions.

The theory begins by stating the force experienced by a particle of unit mass due to a spheroidal body as derived in Appendix A. The relative equation of motion of the planet Mercury with respect to the Sun is

[1] Moulton, F. R. *An Introduction to Celestial Mechanics* (New York: Longmans and Co., 1914), pp. 366-399.

[2] McCuskey, S. W. *Introduction to Celestial Mechanics* (New York: Addison-Wesley, Inc., 1963), pp. 128-143.

obtained by utilizing this force and
the n-body problem.   The result of
this equation yields a mutual perturbative function relating to Mercury
the effects of the spheroidal Sun and
the remaining eight planets.

Each of the planetary orbits is
fully described in terms of its orbital elements*. Moreover, the problem
addressed by this report rests in the
disagreements between predictions made
by Classical Perturbation Theory and
the actual observations of Mercury's
orbital elements. Therefore, the aim
is to account for these discordances
by obtaining the time rates of change
of Mercury's orbital elements as a
result of the perturbation function
when the evaluation of the equation of
motion is taken at both aphelion and
perihelion.

* The orbital elements of a planet are:
the semi-major axis, the eccentricity, the
inclination angle, the argument of the perihelion, the argument of the ascending node,
and the time of perihelion passage. They are
designated by a, e, i, ω, Ω, and T
respectively.

II.

A LITTLE HISTORY

Of all the planetary motions of the solar system, the motion of Mercury around the Sun has been one of the principal problems addressed by Celestial Mechanics. The reason for this lies in the substantial discrepancy between the observed and calculated rates of precession of its orbit. This problem began in 1627 when Kepler calculated mathematical tables from his three laws of planetary motion. His computations predicted to within 5 hours accuracy the transition of Mercury across the solar disk for the year 1631. In the middle of the 19th century Leverrier improved Kepler's mathematical tables by basing his research upon a series of meridianal observations of the planet. He calculated the periodic perturbations on the path of Mercury around the Sun caused by each of the other six planets known at the time, i.e., by Venus, the Earth, Mars, Jupiter, Saturn, and Uranus. Leverrier's tables predicted anterior positions of Mercury in the heavens. These positions were then compared with observed locations of the planet in the past. The deviation between known locations of Mercury and predicted positions was less than 11" of arc.

In spite of this minute deviation, Leverrier was not satisfied with
his tables. He recalculated them a
second time taking Mercury's transit
of November 1697 through the solar
edge as a primary observation and the
November 1848 transit as a final
measurement. This recalculation
included more than 150 years of observational data. After a detailed comparison of the new tables and observation, he found satisfactory agreement
among the majority of the measurements. However, five May transits
observed in the years from 1753 to
1845 presented large discordances when
matched with his modified tables. As
a way of justifying these disagreements of transit, Leverrier suggested
that an unknown mechanism could gener-
ate a series of combined orbital
motions. Nevertheless, to account for
these combined motions, corrections
would have to be made on Mercury's
orbital elements. These would have to
be such as to nearly minimize each
other during the November transits,
and their effects would have to be
additive during the May transits.
Leverrier deduced that the greatest
correction would have to be made for
Mercury's perihelion motion. This
correction had to account for 38" arc
during each century elapsed.

To explain the needed corrections
proposed by Leverrier, astronomers
suggested the idea of imprecision in

the masses of the other planets.
Leverrier found that if the imprecision of masses were true, it would
mean that the mass of Venus would have
to be increased by a factor of 1/7
over its then known value[3] in order to
account for the excess motion of Mercury's perihelion. Since the mass of
Venus was known with a certain degree
of accuracy, an increase in mass of
1/7 could not be accepted. The case of
the shift had to be sought elsewhere.

The possible existence of a
planet, or a series of planetoids,
orbiting between the Sun and Mercury,
was another hypothesis proposed as a
way to justify this excess motion of
the perihelion. This hypothetical
planet would have to act so as to
fully account for the yet-unexplained
perihelion movement. The idea of a
planet moving between the Sun and
Mercury sounded so much more reasonable than the imprecision of masses
that astronomers gave it the name of
Vulcan, even though it was still hypothetical. However, Vulcan remained as
such because during several solar
eclipses astronomers scanned the solar
disk in great detail and never found a
planet, not even a small group of

[3] See Poor, C. L. *Gravitation vs.
Relativity* (New York: G. P. Putnam and Sons,
1925), p. 164.

planetoids.

It must be mentioned that when Leverrier developed his mathematical tables, he only considered the two interior contacts made by Mercury along its passage through the solar disk. When Newcomb heard of this in Washington he recalculated Leverrier's tables with the inclusion of external contacts. The inclusion of the external contacts in the calculations made the analysis painstaking since the position of Mercury had to be taken against the glare produced by the Sun, and the problem at hand required the highest precision possible.

Newcomb's calculations confirmed the discordances among the theoretical and observed motion of Mercury's perihelion, discovered earlier by Leverrier, but by considering the two extra contacts he found the excess in the computed motion to be slightly larger than that predicted by Leverrier by 3.6" arc/century.

To cover for all possibilities for justification of the excess motion, Newcomb made an exhaustive redetermination of the planetary masses, and in doing so he found that they did not contribute either in favor of or against the excess precession of Mercury's orbit. Oddly enough, from his computation tables, Newcomb found a few more anomalies in the other orbital elements of the

planet, in addition to that of the
perihelion, that could not be
explained from Newtonian Mechanics.

The situation later turned worse
with the discovery of two more dis-
crepancies in the orbits of Venus and
Mars. With the former planet the
discrepancy was encountered in the
motion of its line of nodes and with
the latter, the movement of its perihelion.

Thus, the investigations performed independently by Leverrier and
Newcomb clearly exposed the existence
of irregularities in the motions of
Mercury, Venus, and Mars, with the
conclusion that Newtonian Mechanics
was not sufficient to justify them.

In regard to Mercury, its excess
of orbital precession could not be
corrected by any means without modify-
ing the remaining orbital elements.

Although Newcomb's angular mea-
sure for the motion of Mercury's
perihelion was found to be about 41.6"
arc/century, it could not be taken as
an exact value. Newcomb himself gave
a ± 1.5" arc/century as a probable
error in the observations.

Thanks to Leverrier and Newcomb,
however, the excess motion of the
orbit of Mercury proved to be real.
The problem was then to find a physical phenomenon that could satisfac-
torily explain such an oddity without

introducing new complications in the motions of the other planets.

Aside from the hypothetical Vulcan, astronomers considered several other possibilities as answers. The most interesting one was exposed by the same Newcomb[4], who was the first to consider an ovoidal shape of the Sun as a probable factor for inducing a rotation on the orbit of Mercury. But the reason for accepting the structure of the Sun as non-spherical was based mostly on Poor[5],[6], who in a series of two papers published in 1905, explained a deformed solar shape from observations made by Rutherfurd at Columbia University. After comparing these with solar measurements obtained by Auwers[7] in 1891, Poor concluded that there was indeed a difference in the relative sizes of the equatorial and polar radii of the Sun, and this could possibly affect the motion of Mercury.

[4] Newcomb, Simon. *Elements of the Four Inner Planets* (Government printing office, Washington, D.C., 1895).

[5] Poor, C. L. "The Figure of the Sun," *Astrophysical Journal*, Vol. 22, 1905, pp. 103-114.

[6]*Ibid.*, pp. 305-317.

[7] Auwers, A. *Astron. Nachr.*, Vol. 128, 1891, p. 367.

III.

POSITION OF RELATIVITY THEORY

Following his 1905 Special Theory of Relativity paper, which furnished a justification for the null results given by Michelson and Morley in regard to the absence of an absolute frame of reference, Albert Einstein in 1916, proposed a more sophisticated but general conjecture which he entitled "The Foundation of The General Theory of Relativity." It came about from the desire to describe the behavior of the Cosmos from a mathematical viewpoint, in an attempt to unify in a compact fashion, the four natural forces currently known to man.

As it is well known, the Special Theory of Relativity states the speed of light c as the ultimate velocity and concludes that a body in motion is subject to several "physical" effects if observed from an inertial reference frame at rest relative to the moving body. Being observed from the rest frame, the body will vary in size and shape while in motion. Its time will also change with place and velocity. Still, this theory is restricted to situations in which the relative velocities are maintained constant and unidirectionally oriented.

To avoid such restriction, the General Theory of Relativity includes systems with accelerated motions, hence covering all possible behaviors in the Universe at large. A clear evidence of accelerated systems lies in the curved motions of the planets in the Solar System due to their mutual gravitational attractions. This general theory covers the same fundamental concepts as Special Relativity by reassuring that time and space are no longer independent definitions. Both are now jointly related in what is known as the time-space continuum, thus depending on the motion of the body. For example, as time slows down, space appears to shrink, and vice versa.

The general laws of nature, in Relativity Theory, are expressed by equations which are valid independent of the reference frames taken. To Einstein, gravitation seems to be a characteristic of a particular location in space, rather than an effect produced by a body occupying that point at a definite instant. Hence, what must be considered is not the motion of particular objects in space, but the peculiarity of space itself, as affected by the presence of matter. All this implies that the spatial geometry is not uniform throughout, in spite of a nearby gravitational field, but is determined by the presence of matter.

From the mathematics of the theory, space appears to bend in the vicinity of an attractive field in proportion to the magnitude of the mass density producing the field. The bending of space suggests a geometry different from that of Euclidean space. Therefore, since light travels through space and since space is curved in the presence of matter, it is concluded that light also bends when it passes near a concentration of mass. In general, the more dense the mass, the more curved is the nearby space, and light deflects more. One can express this in common language by saying that matter warps space and this warp creates the motion of matter.

Relativists claim that four observations definitely support the theory. One is the velocity-of-light experiment performed by Michelson and Morley in 1887. Another is the shift detection in the spectrum lines of stars. The third observation is the motion of Mercury's perihelion, and the last one is the deflection of electromagnetic waves in the gravitational field of the Sun, as concluded by the solar eclipse of the year 1919. Since the discovery of the precession of Mercury's orbit by Leverrier and Newcomb, no satisfactory justification has been given to such anomaly. However, the equations in the theory of General Relativity seem to explain the nature of this irregularity without requiring additional components.

The difference between Celestial Mechanics and General Relativity is that the latter uses a modified force law which includes terms involving the ratio of the velocity of a particular moving body to that of the speed of light. In algebraic form, the force is expressed as[8]

$$F_g = \frac{GMm}{r^2} \left(1 + 3\,\frac{v^2}{c^2}\right), \tag{1}$$

where v is the velocity tangential to the path of the planet, always perpendicular to the radius vector of its orbit; c is the speed of light; M and m are the masses of the central body and of the planet respectively; r is the line joining the centers of both bodies, and G is the gravitational constant.

By using proper values for the masses of the Sun and Mercury, and by knowing the perpendicular component of motion in Mercury's translation, the value obtained from (1) will give the necessary force to produce a precession of 43" arc/century in the orbit of the planet. Actually, the General Theory of Relativity attributes the

[8]Ramsey, A. S. *Dynamics, Part I* (Cambridge, 1943), pp. 175-176.

orbit of Mercury and its orbital precession to the curvature of space due to the presence of the nearby Sun, not to a non-central force field. Mathematically, however, this comes about from the complementarity of the General theory and the Lorentz transformations. Therefore, any kind of motion of the planets deviates slightly from fixed elliptical paths. Thus, the orbit as a whole tends to rotate about one of its foci.

IV.

A CRITIQUE OF GENERAL RELATIVITY

In addition to Mercury's perihelion precession, General Relativity predicts similar precessions for the orbits of the other planets. This is done by using the same mathematical process as in the case of Mercury, but by employing pertinent data from each individual planet.

The theory, as said before, attributes these precessions to a relativistic non-Euclidean geometry of matter on space, and states the anomalies to be independent of the mutual attractions of the planets on one another.

According to Poor[9], the estimated excess precession values given by Relativity for the first four orbits of the solar system are those of table 1.

Perihelion motion		
Mercury:	+ 43"	arc per century
Venus:	+ 8.6"	arc per century
Earth:	+ 3.8"	arc per century
Mars:	+ 1.3"	arc per century

Table 1.

[9] See footnote 3, p. 191.

For the same configuration, Newcomb, utilizing his tables and observational data, found the following values for the discrepancy in the rotation of each of the first four orbits. As quoted from Poor[10], table 2 shows the discrepancy values* found by Newcomb.

Perihelion motion

Mercury:	+41.6" ± 1.5"	arc per century
Venus:	−7.3" ± 22.3"	arc per century
Earth:	+5.9" ± 5.6"	arc per century
Mars:	+8.1" ± 2.6"	arc per century

Table 2.

[10] *Ibid.*

* The second numerical column in table 2 gives the probable uncertainty of observation.

By comparing Newcomb's computed values with the ones estimated from Relativity, a significant amount of disagreement is noticed. The only relativistic calculation which agrees with the observed values concerns the rotation of the orbit of Mercury. But still, as the reader can verify, the discrepancy between both measurements is substantial. All other estimations deviate considerably from their correspondent observations.

As referred to earlier in this paper, aside from the perihelia motion of the planets, Newcomb, in the late 19th century, testified to several other disturbances in the solar family. Two of these were the variation in the eccentricity of Mercury and the motion of the line of nodes of Venus. In this regard, General Relativity becomes useless since it is unable to even explain in mathematical terminology both the secular distortions in Mercury's eccentricity and the nodes of Venus. These distortions amount to the exceeded values of -0.88" and +10.2" arc per century, respectively. Relativity does not predict the excess perturbations in the orbits of the planets other than those of the motions of their perihelia. Still, of all the secular perturbations observed, General Relativity agrees only with the motion of the orbit of Mercury. It fails to account for more or less than this perturbation since there exists no flexibility in the equations of Einstein, no uncertainty in the calculations, and no room for compensation.

Take the case of Venus for example. Relativity estimates an excess orbital rotation rate of 8.6" arc per century with a forward motion. Yet, the computed excess rotational rate was found to be -7.3" arc per century with a retrograde motion!

It is thought here that the forward rotation of Mercury's orbit from Relativity is a mere coincidence since Relativity theory predicts the correct anomaly for this sole case. Amazingly enough, however, relativists stress this coincidence as one of the four conclusive proofs for the curved geometry of space and General Relativity.

The motion of the other perihelia has not been the only fallacy found in Einstein's theories. The alleged deflection of light in the vicinity of a massive body has proven to be inconclusive.[11] The shifting of the lines in the electromagnetic spectrum of stars, a presumed test of General Relativity and believed to occur because of the uniform recession of the

---

[11] See Poor, footnote 3, pp. 197, 226.

stars, is also challenged by Dingle[12] and Slusher.[13] They suggest that the shift in the lines is merely a Doppler effect of velocities of stars existing now, and not ages ago, as is inferred by relativists. Even the second postulate of Special Relativity has been proven wrong by Kantor[14] in one of his light interferometer experiments.

In response to the avalanche of fallacies found in the present system, some alternatives have been proposed, such as the ones formulated by Barnes,

[12] Dingle, H. "The Doppler Effect and The Foundations of Physics" (I) and (II). *British Journal for the Philosophy of Science XI*, Vol. 41, 1960, pp. 11-31, and Vol. 42, 1960, pp. 113-129.

[13] Slusher, H. S. "Cosmology and Einstein's Postulate of Relativity." *Creation Research Society Quarterly*, Vol. 17, No. 3, Dec. 1980, pp. 146-147.

[14] Kantor, W. *Relativistic Propagation of Light* (Lawrence, Kansas: Coronado Press, 1976), pp. 142-149.

Pemper, and Armstrong[15], and Waldron[16], which show no deviation from Euclidean geometry (and logic), and agree to a large extent with the facts, without the need to confront inconsistencies among their arguments.

[15] Barnes, T. G., Pemper, R. R., and Armstrong, H. L. "A Classical Foundation for Electrodynamics." *Creation Research Society Quarterly*, Vol. 14, No. 1, June 1977, pp. 38-45.

[16] Waldron, R. A. *The Wave and Ballistic Theories of Light-A Critical Review* (Frederick Muller Limited, 1977), pp. 125-165.

V.

THE THEORY

A.   General Basis For The Theory

Because any deviation from the Keplerian motion of two spherically homogeneous bodies is considered a perturbation, the solar system is constantly subject to two perturbative sources. One of these sources is the mutual force interaction of the planets on each other. The other source is the gravitational attraction produced by a deformed Sun. At the current level of mathematical development, it is impossible to formulate an exact solution for the motions of a multiple-body system; the kinematics are only approximated from Kepler's three laws of planetary motion and from Newtonian Gravitation. Hence, a different approach must be searched for. This different approach is known as the n-body problem. In spite of being just another approximation, its pursuit is to describe the perturbative effects in a system composed of three or more bodies in order that their motions may be described with more precision than from Kepler's laws.

After Poor's conclusion[17] of a

---

[17] See footnotes 5 and 6.

3.6 ± .23 x 10^{-5} solar oblateness value*, the Sun has been the motive of vast analyzation by several other investigators. The most important studies have been done by Dicke and Goldenberg[18], and by Gilvarry and Sturrock[19]. Dicke and Goldenberg, for example, found a solar oblateness of 5.0 ± 0.7 x 10^{-5} during the year of 1967. This value was reached as a conclusion after having exhaustively analyzed all data at hand. On the other hand, Gilvarry and Sturrock suggested that the solar oblateness given by Dicke might be due to a rapidly spinning solar core.

The theory of this monograph thus, is based upon a gravitational force produced by the Sun. It accepts Poor's, and Dicke and Goldenberg's

* The oblateness value is just the difference taken between the equatorial and polar solar radii, divided by the equatorial one.

[18] Dicke, R. H. and Goldenberg, H. M. "Solar Oblateness and General Relativity," *Physical Review Letters*, Vol. 18, Feb. 27, 1967, pp. 313-316.

[19] Gilvarry, J. J. and Sturrock, P. A. "Solar Oblateness and the Perihelion Advances of Planets," *Nature*, Vol. 216, Dec. 1967, pp. 1283-1285.

conclusions of a solar oblateness. Furthermore, it assumes an invariant solid solar shape.

The sizes of the planets are small when compared with that of the Sun. Hence, for all practical purposes they will be considered as small homogeneous spheres. The system corresponds to interactions between the gravitational fields originating from an oblate massive body and minute spheres. Under this type of situation the forces between the Sun and the planets no longer act along the lines joining their respective centers and is said to be a non-central force.

Due to the assumption of a solid rotating shape, the Sun will possess equatorial and axial moments of inertia. Taking the solar rotational axis as the z' orientation for simplicity, the force experienced by a particle of unit mass located at a distance r' from the center of an oblate Sun of mass m_1 is given by*

$$\vec{F}' = \frac{-Gm_1\vec{r}'}{r'^3} - \frac{3G\Delta I}{2r'^5}\left(1 - \frac{5z'^2}{r'^2}\right)\vec{r}' - \frac{3G\Delta I}{r'^5}\vec{z}', \quad (2a)$$

where $\Delta I = \frac{m_1}{5}\left(R_e^2 - R_p^2\right)$,[+] and G is the gravitational constant.

    * See Appendix A for a derivation of (2a).

    [+] R_e and R_p represent the equatorial and polar solar radii respectively.

The first term on the right hand side of (2a) is just the expression for an ordinary gravitational central force. The other two terms arise as contributions due to a perturbative force from the solar oblateness.

B.   Motion of a Planet Relative to the Sun Under the Presence of $n-2$ Bodies

Let $i = 2, 3, 4, \ldots, n$, and consider now $n-1$ planets having masses m_i located at distances r_{1i}' from the center of the Sun. In this case the only force strong enough to affect their motions is that of Gravitation. Fig. 1 illustrates the situation as viewed from a fixed rectangular coordinate system (x_o, y_o, z_o) parallel to the solar (x', y', z') frame.

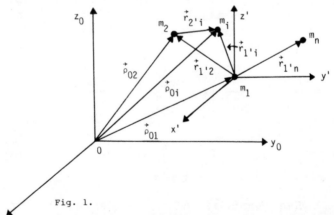

Fig. 1.

Geometrical representation of a system of n bodies with their related vectorial distances. The plane x'y' is the solar equatorial plane. x' is oriented towards the vernal equinox.

Because $m_1 \gg m_i$, the gravitational attraction between two planets is a central force given by

$$\vec{F}'_{ij} = -\frac{Gm_i m_j}{r'^3_{ij}} \vec{r}'_{ij}, \quad i, j = 2, 3, 4, \ldots n \ (i \neq j), \qquad (2b)$$

as calculated from the (x', y', z') reference frame. Furthermore, the force exerted by m_1 on either of the planets is that given by the right-hand side of (2a) times m_i. Note that because the bodies are in constant motion with respect to one another, they must obey Newton's non-relativistic equation of motion*

$$\vec{F} = \frac{d\vec{p}}{dt}, \qquad (3)$$

if $\vec{p}$ is the momentum of the body.

Expressing the length vectors $\vec{\rho}_{o1}$, $\vec{\rho}_{oi}$ in terms of velocity, the following identities are obtained.

* The velocity v_i of the planet m_i is small compared with the speed of light.

$$\vec{v}_{o1} = \frac{d\vec{\rho}_{o1}}{dt}, \quad \vec{v}_{oi} = \frac{d\vec{\rho}_{oi}}{dt}, \quad i = 2, \, 3, \, 4, \, \ldots, \, n.$$

With the aid of these identities the individual equations of motion for the n bodies are:

$$\frac{m_1}{G} \frac{d\vec{v}_{o1}}{dt} = \sum_{i=2}^{n} \left[\frac{m_i m_1}{r_{1i}'^3} \vec{r}_{1i}' + \frac{3\Delta I m_i}{2r_{1i}'^5} \left(1 - \frac{5z_{1i}'^2}{r_{1i}'^2}\right) \vec{r}_{1i}' + \frac{3\Delta I}{r_{1i}'^5} m_i \vec{z}_{1i}' \right], \qquad (4a)$$

$$\frac{m_i}{G} \frac{d\vec{v}_{oi}}{dt} = \sum_{j=2}^{n} \frac{m_i m_j}{r_{ij}'^3} \vec{r}_{ij}' + \frac{m_i m_1}{r_{il}'^3} \vec{r}_{il}' + \frac{3 m_i \Delta I}{2 r_{il}'^5} \left(1 - \frac{5z_{il}'^2}{r_{il}'^2}\right) \vec{r}_{il}'$$

$$+ \frac{3 m_i \Delta I}{r_{il}'^5} \vec{z}_{il}', \quad i, \, j = 2, \, 3, \, 4, \, \ldots, \, n \; (i \neq j), \qquad (4b)$$

where $\vec{r}_{1i}'$ is the vector from m_1 to m_i and $\vec{r}_{ij}'$ is the vector from m_i to m_j. Observe that $\vec{r}_{1i}' = -\vec{r}_{il}'$ and $\vec{r}_{ij}' = -\vec{r}_{ji}'$; also $\vec{z}_{1i}' = -\vec{z}_{il}'$.

For $i = 2$, (4a) & (4b) become

$$\frac{m_1}{G} \frac{d\vec{V}_{o1}}{dt} = \frac{m_1 m_2}{r_{12}'^3} \vec{r}_{12}' + \frac{3\Delta I m_2}{2r_{12}'^5} \left(1 - \frac{5z_{12}'^2}{r_{12}'^2}\right) \vec{r}_{12}' + \frac{3\Delta I}{r_{12}'^5} m_2 \vec{z}_{12}'$$

$$+ \sum_{j=3}^{n} \left[\frac{m_1 m_j}{r_{1j}'^3} \vec{r}_{1j}' + \frac{3\Delta I m_j}{2r_{1j}'^5} \left(1 - \frac{5z_{1j}'^2}{r_{1j}'^2}\right) \vec{r}_{1j}' + \frac{3\Delta I}{r_{1j}'^5} m_j \vec{z}_{1j}'\right], \qquad (4c)$$

$$\frac{m_2}{G} \frac{d\vec{V}_{o2}}{dt} = \sum_{j=3}^{n} \frac{m_2 m_j}{r_{2j}'^3} \vec{r}_{2j}' + \frac{m_2 m_1}{r_{21}'^3} \vec{r}_{21}' + \frac{3m_2 \Delta I}{2r_{21}'^5} \left(1 - \frac{5z_{21}'^2}{r_{21}'^2}\right) \vec{r}_{21}'$$

$$+ \frac{3m_2 \Delta I}{r_{21}'^5} \vec{z}_{21}'. \qquad (4d)$$

Because the vectors $\vec{r}_{1i}'$ and $\vec{r}_{ij}'$ are measured from the origin of the (x', y', z') coordinate system subtraction of (4c) from (4d) yields a motion of m_2 relative to m_1, since $\vec{\rho}_{o2} - \vec{\rho}_{o1} = \vec{r}_{12}'$. Therefore,

$$\frac{1}{G} \frac{d}{dt} (\vec{V}_{o2} - \vec{V}_{o1}) = \sum_{j=3}^{n} \frac{m_j}{r_{2j}'^3} \vec{r}_{2j}' - \frac{\mu}{r_{12}'^3} \vec{r}_{12}' - \frac{3\mu}{2r_{12}'^5} J \left(1 - \frac{5z_{12}'^2}{r_{12}'^2}\right) \vec{r}_{12}'$$

$$-\frac{3\mu J}{r_{12}^{'5}}\vec{z}_{12}^{\,'} - \sum_{j=3}^{n}\left[\frac{m_j}{r_{1j}^{'3}}\vec{r}_{1j}^{\,'} + \frac{3Jm_j}{2r_{1j}^{'5}}\left(1 - \frac{5z_{1j}^{'2}}{r_{1j}^{'2}}\right)\vec{r}_{1j}^{\,'}\right.$$

$$\left.+\frac{3Jm_j}{r_{1j}^{'5}}\vec{z}_{1j}^{\,'}\right] = \frac{1}{G}\frac{d^2\vec{r}_{12}^{\,'}}{dt^2}, \tag{5}$$

where $J = \Delta I/m_1$ and $\mu = m_1 + m_2$.

Fig. 2 shows the isolation of the solar $(x', \dot{y}', z')$ frame from the (x_o, y_o, z_o) rest coordinate system.

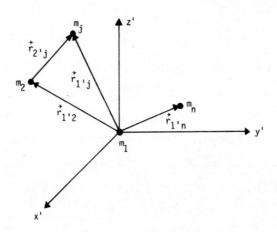

Fig. 2. A system of n bodies being considered from the (x', y', z') frame. x' points towards the vernal equinox and z' is the solar rotational axis. Notice that $\vec{r}_{12}^{\,'} < \vec{r}_{13}^{\,'} < \vec{r}_{14}^{\,'} \ldots < \vec{r}_{1j}^{\,'} < \ldots < \vec{r}_{1n}^{\,'}$.

It is known from astronomy and celestial mechanics that the equatorial plane of the sun is inclined at an angle γ from the ecliptic plane.* According to Dicke[20], this angle has a value of 7.25°. Thus, because every pertinent data taken about the motion of the heavenly bodies is always measured relative to the ecliptic, the terminology so far used in this paper must be transformed from solar coordinates to ecliptic coordinates. In doing this, a mathematical tool known as infinitesimal rotations[21] is used. If (x, y, z) represents the ecliptic coordinates, the transformation is found by letting

$$x = x'$$

$$y = y' \cos \gamma - z' \sin \gamma$$

$$z = y' \sin \gamma + z' \cos \gamma.$$

* The ecliptic plane is the earth's orbital plane.

[20] Dicke, R. H. "The Oblateness of the Sun and Relativity," *Science*, Vol. 184, April 26, 1974, p. 420.

[21] See Goldstein, H. *Classical Mechanics* (New York: Addison-Wesley Inc., 1950), pp. 124, 127.

The x axis remains invariant under the transformation and it points towards the vernal equinox.   Fig. 3 illustrates the rotation angle between both the solar and the ecliptic frames.

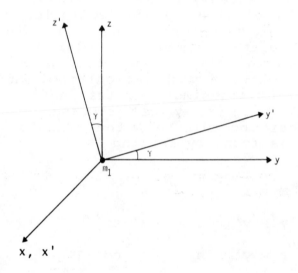

Fig. 3 Orientation of the ecliptic reference frame relative to the solar coordinate system. Note that m_1 is located at the origin of both frames.

For the transformation, the vector relationships given in Fig. 2 change only in direction. Their magnitudes remain invariant. Hence

$$\vec{r}_{1j}' \rightarrow \vec{r}_{1j}$$

$$\vec{r}_{ij}' \rightarrow \vec{r}_{ij}, \qquad (6)$$

such that

$$\vec{r}_{12} < \vec{r}_{13} < \ldots < \vec{r}_{1j} < \ldots < \vec{r}_{1n}.$$

Conversion of (5) to ecliptic coordinates gives

$$\frac{1}{G}\frac{d^2\vec{r}_{12}}{dt^2} = \sum_{j=3}^{n} \left[\frac{m_j}{r_{2j}^3}\vec{r}_{2j} - \frac{m_j}{r_{1j}^3}\vec{r}_{1j} - \frac{3Jm_j}{2r_{1j}^5}\left(1 - \frac{5z_{1j}^2}{r_{1j}^2}\right)\vec{r}_{1j} \right.$$

$$\left. - \frac{3Jm_j}{r_{1j}^5}\vec{z}_{1j} \right] - \frac{\mu}{r_{12}^3}\vec{r}_{12} - \frac{3\mu J}{2r_{12}^5}\left(1 - \frac{5z_{12}^2}{r_{12}^2}\right)\vec{r}_{12} - \frac{3\mu J}{r_{12}^5}\vec{z}_{12}. \qquad (7)$$

From (7), the x_{12} component of $d^2\vec{r}_{12}/dt^2$ is

$$\frac{d^2 x_{12}}{G dt^2} = \sum_{j=3}^{n} \left[\frac{m_j(x_{1j} - x_{12})}{r_{2j}^3} - \frac{m_j}{r_{1j}^3} x_{1j} - \frac{3Jm_j}{2r_{1j}^5} \left(1 - \frac{5z_{1j}^2}{r_{1j}^2}\right) x_{1j} \right]$$

$$- \frac{\mu}{r_{12}^3} x_{12} - \frac{3\mu J}{2r_{12}^5} \left(1 - \frac{5z_{12}^2}{r_{12}^2}\right) x_{12}, \qquad (8)$$

Since the $\hat{x}$ and $\hat{z}$ orientations are always mutually orthogonal.*

Letting $K = 3J/2$ and re-arranging the terms in (8) yields

$$\frac{1}{G} \frac{d^2 x_{12}}{dt^2} = \sum_{j=3}^{n} \left[\frac{m_j(x_{1j} - x_{12})}{r_{2j}^3} - \frac{m_j x_{1j}}{r_{1j}^3} \left(1 + \frac{K}{r_{1j}^2} \left(1 - \frac{5z_{1j}^2}{r_{1j}^2}\right)\right) \right]$$

$$- \frac{\mu}{r_{12}^3} x_{12} \left[1 + \frac{K}{r_{12}^2} \left(1 - \frac{5z_{12}^2}{r_{12}^2}\right)\right]. \qquad (9)$$

From (6), the magnitude of the vector $\vec{r}_{2j}$ is

$$\left|\vec{r}_{2j}\right| = \left[(x_{1j} - x_{12})^2 + (y_{1j} - y_{12})^2 + (z_{1j} - z_{12})^2\right]^{1/2} \qquad (10a)$$

* Where $\hat{x} = \vec{x}_{12}/x_{12}$ and $\hat{z} = \vec{z}_{12}/z_{12}$.

from which the following relations are deduced:

$$\frac{\partial}{\partial x_{12}} \left(\frac{1}{r_{2j}}\right) = \frac{x_{1j} - x_{12}}{r_{2j}^3}; \quad \frac{\partial}{\partial y_{12}} \left(\frac{1}{r_{2j}}\right) = \frac{y_{1j} - y_{12}}{r_{2j}^3};$$

$$\frac{\partial}{\partial z_{12}} \left(\frac{1}{r_{2j}}\right) = \frac{z_{1j} - z_{12}}{r_{2j}^3}. \tag{10b}$$

But since the location of m_2 is independent of the position of any of the m_j's one can say that

$$\frac{\partial}{\partial x_{12}} (\vec{r}_{12} \cdot \vec{r}_{1j}) = x_{1j},$$

$$\frac{\partial}{\partial y_{12}} (\vec{r}_{12} \cdot \vec{r}_{1j}) = y_{1j}, \tag{10c}$$

$$\frac{\partial}{\partial z_{12}} (\vec{r}_{12} \cdot \vec{r}_{1j}) = z_{1j},$$

Combining the individual differentiations of both (10b) and (10c) multiplied by $1/r_{1j}^3$ yields:

$$\frac{\partial}{\partial x_{12}} \left(\frac{1}{r_{2j}}\right) - \frac{\partial}{\partial x_{12}} \left(\frac{\vec{r}_{12} \cdot \vec{r}_{1j}}{r_{1j}^3}\right) = \frac{\partial}{\partial x_{12}} H_j,$$

$$\frac{\partial}{\partial y_{12}} \left(\frac{1}{r_{2j}}\right) - \frac{\partial}{\partial y_{12}} \left(\frac{\vec{r}_{12} \cdot \vec{r}_{1j}}{r_{1j}^3}\right) = \frac{\partial}{\partial y_{12}} H_j, \text{ and}$$

$$\frac{\partial}{\partial z_{12}} \left(\frac{1}{r_{2j}}\right) - \frac{\partial}{\partial z_{12}} \left(\frac{\vec{r}_{12} \cdot \vec{r}_{1j}}{r_{1j}^3}\right) = \frac{\partial}{\partial z_{12}} H_j.$$

Thus, for m_2 relative to m_1,

$$H_j = \frac{1}{r_{2j}} - \frac{\vec{r}_{12} \cdot \vec{r}_{1j}}{r_{1j}^3}. \tag{10d}$$

H_j in (10d) is defined as a perturbative function of position.[22]

Using (10d), (9) can be rewritten as

$$\frac{d^2 x_{12}}{G dt^2} = \sum_{j=3}^{n} m_j \left[\frac{\partial H_j}{\partial x_{12}} - \frac{K}{r_{1j}^2} \left(1 - \frac{5 z_{1j}^2}{r_{1j}^2}\right) \frac{\partial}{\partial x_{12}} \left(\frac{1}{r_{2j}} - H_j\right)\right]$$

$$- \mu \left[\frac{x_{12}}{r_{12}^3} + \frac{K x_{12}}{r_{12}^5} \left(1 - \frac{5 z_{12}^2}{r_{12}^2}\right)\right]. \tag{11a}$$

[22] See McCuskey, S. W., op. cit., p. 102.

Following the same algebraic process so far, similar expressions for the y_{12} and z_{12} components of the acceleration of m_2 relative to m_1 are obtained. These are

$$\frac{d^2 y_{12}}{G dt^2} = \sum_{j=3}^{n} m_j \left[\frac{\partial H_j}{\partial y_{12}} - \frac{K}{r_{1j}^2} \left(1 - \frac{5z_{1j}^2}{r_{1j}^2}\right) \frac{\partial}{\partial y_{12}} \left(\frac{1}{r_{2j}} - H_j\right) \right]$$

$$- \mu \left[\frac{y_{12}}{r_{12}^3} + \frac{K y_{12}}{r_{12}^5} \left(1 - \frac{5z_{12}^2}{r_{12}^2}\right) \right], \tag{11b}$$

$$\frac{d^2 z_{12}}{G dt^2} = \sum_{j=3}^{n} m_j \left[\frac{\partial H_j}{\partial z_{12}} - \frac{K}{r_{1j}^2} \left(1 - \frac{5z_{1j}^2}{r_{1j}^2}\right) \frac{\partial}{\partial z_{12}} \left(\frac{1}{r_{2j}} - H_j\right) - \frac{2Kz_{1j}}{r_{1j}^5} \right]$$

$$- \mu \left[\frac{z_{12}}{r_{12}^3} + \frac{K z_{12}}{r_{12}^5} \left(1 - \frac{5z_{12}^2}{r_{12}^5}\right) + \frac{2K z_{12}}{r_{12}^5} \right]. \tag{11c}$$

Eq. (11c) contains two extra terms which are acceleration components attributed to the solar oblateness.

In vector notation (11a) through (11c) yield

$$\frac{1}{G}\frac{d^2\vec{r}_{12}}{dt^2} = \sum_{j=3}^{n} m_j \left[\vec{\nabla}_{12}H_j - \frac{K}{r_{1j}^2}\left(1 - \frac{5z_{1j}^2}{r_{1j}^2}\right)\vec{\nabla}_{12}\left(\frac{1}{r_{2j}} - H_j\right) - \frac{2K\vec{z}_{1j}}{r_{1j}^5}\right]$$

$$- \mu \left[\frac{\vec{r}_{12}}{r_{12}^3} + \frac{K}{r_{12}^5}\left(1 - \frac{5z_{12}^2}{r_{12}^2}\right)\vec{r}_{12} + \frac{2K\vec{z}_{12}}{r_{12}^5}\right] \tag{12}$$

if the del operator $\vec{\nabla}_{12} \equiv (\partial/\partial x_{12})\,\hat{x} + (\partial/\partial y_{12})\,\hat{y} + (\partial/\partial z_{12})\,\hat{z}$. Hence (12) determines the motion of m_2 relative to m_1, as viewed from the ecliptic plane when the effects caused by $\sum_{j}^{n} m_j$ and the solar oblateness are included.

C. Acceleration $d^2\vec{r}_{12}/dt^2$ and the Orbital Elements of m_2

Suppose it is desired to express the coordinates of m_2 in terms of m_2's orbital elements and time. If α_1, α_2, . . ., α_6 represent the elments of the orbit described by m_2, the orthogonal equations for the coordinates may be written as

$$x_{12} = x_{12}(\alpha_1, \alpha_2, . . ., \alpha_6, t),$$

$$y_{12} = y_{12} (\alpha_1, \alpha_2, \ldots, \alpha_6, t),$$

$$z_{12} = z_{12} (\alpha_1, \alpha_2, \ldots, \alpha_6, t). \qquad (13)$$

For the ideal case in which m_1 is spherical and $\overset{n}{\underset{j}{\Sigma}} m_j$ are absent, the orbital motion of m_2 about the Sun m_1 is said to be Keplerian (unperturbed), and the elements describing the orbit remain constant in time. But in reality, the oblateness of m_1 and the presence of $\overset{n}{\underset{j}{\Sigma}} m_j$ must be taken into account. Thus, the motion of m_2 slowly deviates from the hypothetical Keplerian path. And for the true perturbed orbital motion of m_2 the elements are actual functions of time. The problem then is to find the time rates of change of the α's when the perturbation effects just mentioned are included.

The process begins by taking a differentiation of (13) with respect to time in the true motion of m_2 about m_1. This gives

$$\frac{d\vec{r}_{12}}{dt} = \sum_{k=1}^{6} \frac{\partial \vec{r}_{12}}{\partial \alpha_K} \frac{d\alpha_K}{dt} + \frac{\partial \vec{r}_{12}}{\partial t}, \qquad (14)$$

where $\vec{r}_{12}$ is the vector from m_1 to m_2.

Because the orbital elements are slowly changing in time, one can assume the change to be uniform and consider their accelerations as nonexistent. With this, a second time differentiation of (13) yields

$$\frac{d^2\vec{r}_{12}}{dt^2} = \sum_{k=1}^{6} \frac{\partial^2\vec{r}_{12}}{\partial t \partial \alpha_K} \frac{d\alpha_K}{dt} + \frac{\partial^2\vec{r}_{12}}{\partial t^2}. \qquad (15)$$

Substitution of (15) in (12) gives

$$\sum_{k=1}^{6} \frac{\partial^2\vec{r}_{12}}{\partial t \partial \alpha_K} \frac{d\alpha_K}{dt} + \frac{\partial^2\vec{r}_{12}}{\partial t^2} + \frac{G\mu}{r_{12}^3}\left[\vec{r}_{12} + \frac{K\vec{r}_{12}}{r_{12}^2}\left(1 - \frac{5z_{12}^2}{r_{12}^2}\right) + \frac{2Kz\vec{r}_{12}}{r_{12}^2}\right]$$

$$= G\sum_{j=3}^{n} m_j\left[\vec{v}_{12} H_j - \frac{K}{r_{1j}^2}\left(1 - \frac{5z_{1j}^2}{r_{1j}^2}\right)\vec{v}_{12}\left(\frac{1}{r_{2j}} - H_j\right) - \frac{2Kz\vec{r}_{1j}}{r_{1j}^5}\right]. \qquad (16)$$

For the Keplerian motion of m_2 no perturbative forces arise and H_j along with K disappear. And because for this case the elements are invariant with time, from (16) one gets

$$\frac{\partial^2\vec{r}_{12}}{\partial t^2} = -\frac{G\mu\vec{r}_{12}}{r_{12}^3}. \qquad (17)$$

Vector analysis states that the grad-
ient of an algebraic sum equals the
algebraic sum of the gradients.
Therefore, it is valid to write

$$\vec{\nabla}_{12}\left(\frac{1}{r_{2j}} - H_j\right) = \vec{\nabla}_{12}\left(\frac{1}{r_{2j}}\right) - \vec{\nabla}_{12}H_j. \qquad (18)$$

Substitution of (17) and (18) in (16)
yields

$$\sum_{k=1}^{6} \frac{\partial^2 \vec{r}_{12}}{\partial t \partial \alpha_K}\frac{d\alpha_K}{dt} + \frac{G\mu K\vec{r}_{12}}{r_{12}^5}\left(1 - \frac{5z_{12}^2}{r_{12}^2}\right) + 2GK\left[\frac{\mu}{r_{12}^5}\vec{z}_{12} + \sum_{j=3}^{n}\frac{m_j\vec{z}_{1j}}{r_{1j}^5}\right]$$

$$= G\sum_{j=3}^{n} m_j\left\{\left[\left(1 + \frac{K}{r_{1j}^2}\left(1 - \frac{5z_{1j}^2}{r_{1j}^2}\right)\right)\vec{\nabla}_{12}H_j - \frac{K}{r_{1j}^2}\left(1 - \frac{5z_{1j}^2}{r_{1j}^2}\right)\vec{\nabla}_{12}\left(\frac{1}{r_{2j}}\right)\right]\right\}.(19)$$

Letting

$$L_1 = \frac{G\mu K}{r_{12}^4}\left(1 - \frac{5z_{12}^2}{r_{12}^2}\right),$$

$$L_2 = \frac{2GK\mu z_{12}}{r_{12}^5}, \qquad (20)$$

$$L_{3j} = \frac{2GKm_j}{r_{1j}^5} z_{1j},$$

$$L_{4j} = Gm_j \left[1 + \frac{K}{r_{1j}^2} \left(1 - \frac{5z_{1j}^2}{r_{1j}^2} \right) \right],$$

$$L_{5j} = \frac{GKm_j}{r_{1j}^2} \left(1 - \frac{5z_{1j}^2}{r_{1j}^2} \right),$$

(19) reduces to*

$$\sum_{k=1}^{6} \frac{\partial^2 \vec{r}_{12}}{\partial t \partial \alpha_K} \frac{d\alpha_K}{dt} = - L_1 \hat{r}_{12} - \left(L_2 + \sum_{j=3}^{n} L_{3j} \right) \hat{z} + \sum_{j=3}^{n}$$

$$\left[L_{4j} \vec{\nabla}_{12} H_j - L_{5j} \vec{\nabla}_{12} \left(\frac{1}{r_{2j}} \right) \right]. \tag{21}$$

From the paragraph preceeding (14),

* The interested reader will notice that $\vec{z}_{1j}/z_{1j} = \vec{z}_{12}/z_{12} = \hat{z}$.

the time rates of change of the orbital elements are defined such that

$$\sum_{k=1}^{6} \frac{\partial \vec{r}_{12}}{\partial \alpha_K} \frac{d\alpha_K}{dt} = 0 \tag{22}$$

for the unperturbed motion of m_2 around m_1. The expressions (21) and (22) must be solved for $d\alpha_K/dt$. Dotting (21) with $\partial \vec{r}_{12}/\partial \alpha_1$ shows that

$$\sum_{k=1}^{6} \frac{\partial^2 \vec{r}_{12}}{\partial t \partial \alpha_K} \cdot \frac{\partial \vec{r}_{12}}{\partial \alpha_1} \left(\frac{d\alpha_K}{dt}\right) = -L_1 \frac{\partial r_{12}}{\partial \alpha_1} - \left(L_2 + \sum_{j=3}^{n} L_{3j}\right) \frac{\partial z_{12}}{\partial \alpha_1}$$

$$+ \sum_{j=3}^{n} \left[L_{4j} \frac{\partial H_j}{\partial \alpha_1} - L_{5j} \frac{\partial}{\partial \alpha_1} \left(\frac{1}{r_{2j}}\right)\right]. \tag{23}$$

By performing the same operation in (22) with $\partial^2 \vec{r}_{12}/\partial t \partial \alpha_1$ and subtracting this from (23), the total contribution of the perturbations yields

$$\sum_{k=1}^{6} \left(\frac{\partial \vec{r}_{12}}{\partial \alpha_1} \cdot \frac{\partial^2 \vec{r}_{12}}{\partial t \partial \alpha_K} - \frac{\partial \vec{r}_{12}}{\partial \alpha_K} \cdot \frac{\partial \vec{r}_{12}}{\partial t \partial \alpha_1}\right) \frac{d\alpha_K}{dt} = -L_1 \frac{\partial r_{12}}{\partial \alpha_1}$$

$$- (L_2 + \sum_{j=3}^{n} L_{3j}) \frac{\partial z_{12}}{\partial \alpha_1} + \sum_{j=3}^{n} [L_{4j} \frac{\partial H_j}{\partial \alpha_1} - L_{5j} \frac{\partial}{\partial \alpha_1} (\frac{1}{r_{2j}})], \qquad (24)$$

where $1 = 1, 2, \ldots, 6$; $1 \neq K$, since the order of terms in a scalar product is interchangeable.

In classical mechanics the left hand side of (24) is commonly known as Lagrange brackets.[23] Generally, they are expressed as

$$\sum_{k} (\frac{\partial q_K}{\partial u} \frac{\partial p_K}{\partial v} - \frac{\partial p_K}{\partial u} \frac{\partial q_K}{\partial v}) = {}_k [u, v]_{q, p}.$$

Comparing the similarity among the above brackets with the quantities $\vec{r}_{12}$, $\partial \vec{r}_{12} / \partial t$, α_1, α_K, (24) reduces even more so that

$$\sum_{k=1}^{6} [\alpha_1, \alpha_K] \frac{d\alpha_K}{dt} = - L_1 \frac{\partial r_{12}}{\partial \alpha_1} - (L_2 + \sum_{j=3}^{n} L_{3j}) \frac{\partial z_{12}}{\partial \alpha_1}$$

$$+ \sum_{j=3}^{n} [L_{4j} \frac{\partial H_j}{\partial \alpha_1} - L_{5j} \frac{\partial}{\partial \alpha_1} (\frac{1}{r_{2j}})], \ k \neq 1; \ 1 = 1, 2, \ldots, 6. \qquad (25)$$

[23] See Goldstein, H., op. cit., p. 250.

D. Solutions   to   the   Differential
    Expression (25)

The method employed to solve the
Lagrange brackets is the Jacobian
determinant[24] whose general form can
be expressed

$$\frac{\partial(q_k,\ p_K)}{\partial(u,\ v)} = \begin{vmatrix} \dfrac{\partial q_K}{\partial u} & \dfrac{\partial p_K}{\partial u} \\[2ex] \dfrac{\partial q_K}{\partial v} & \dfrac{\partial p_K}{\partial v} \end{vmatrix}$$

Since for every l six k's are
considered, there are apparently
twelve equations that need to be
solved for (25) above. Nevertheless,
one of the properties of Lagrange
brackets is that $[\alpha_l,\ \alpha_K] = -[\alpha_K,\ \alpha_l]$. Hence, the number of equations
to solve reduces to only six.
Independent of the relation that might
exist between the Lagrange brackets
and the right hand side of (25), their
evaluation in terms of the orbital
elements has already been performed by
several authors.[25] The reader is

[24] *Ibid.*, pp. 248, 249.

[25] See for example: Smart, W. M.
Celestial Mechanics (New York: Longmans and
Co., 1953), p. 63. Moulton, F. R., op. cit.,
pp. 391-398. McCuskey, S. W., op. cit.,
p. 141.

requested to consult Appendix B for an understanding of their evaluation from the Jacobian determinants. [In order to avoid loosing track of the theory only the results of that appendix are re-written here, along with their relationship to (25).]

By definition, the α's represent the elements a, e, i, ω, Ω and T of the orbit described by m_2 in its motion around m_1. Combining these elements so as to satisfy the determinants, one finds from Appendix B that

$$\alpha_1 = \Omega, \ \alpha_K = a; \quad [\Omega, \ a] = \tfrac{1}{2} \, \text{nacos} \ i \ \sqrt{1 - e^2},$$

$$\alpha_1 = \omega, \ \alpha_K = a; \quad [\omega, \ a] = \tfrac{1}{2} \ \text{na} \ \sqrt{1 - e^2},$$

$$\alpha_1 = e, \ \alpha_K = \Omega; \quad [e, \ \Omega] = \frac{na^2 e}{\sqrt{1 - e^2}} \cos \ i,$$

$$\alpha_1 = e, \ \alpha_K = \omega; \quad [e, \ \omega] = \frac{na^2 e}{\sqrt{1 - e^2}}, \qquad (26)$$

$$\alpha_1 = i, \ \alpha_K = \Omega; \quad [i, \ \Omega] = na^2 \ \sqrt{1 - e^2} \ \sin \ i,$$

$$\alpha_1 = a, \quad \alpha_K = T; \quad [a, T] = \frac{1}{2}n^2 a,$$

where $n = \dfrac{\sqrt{G(m_1 + m_2)}}{a^{3/2}}$ is the orbital angular speed. All other permissible combinations vanish.

The relation of (26) to (25) yields

$$[\Omega, a]\frac{da}{dt} = \frac{1}{2}na\frac{da}{dt}\cos i \sqrt{1 - e^2} = -L_1 \frac{\partial r_{12}}{\partial \Omega} - (L_2 + \sum_{j=3}^{n} L_{3j}) \frac{\partial z_{12}}{\partial \Omega}$$

$$+ \sum_{j=3}^{n} [L_{4j} \frac{\partial H_j}{\partial \omega} - L_{5j} \frac{\partial}{\partial \omega}(\frac{1}{r_{2j}})], \tag{27}$$

$$[\omega, a]\frac{da}{dt} = \frac{1}{2}na\frac{da}{dt}\sqrt{1 - e^2} = -L_1 \frac{\partial r_{12}}{\partial \omega} - (L_2 + \sum_{j=3}^{n} L_{3j}) \frac{\partial z_{12}}{\partial \omega}$$

$$+ \sum_{j=3}^{n} [L_{4j} \frac{\partial H_j}{\partial \Omega} - L_{5j} \frac{\partial}{\partial \Omega}(\frac{1}{r_{2j}})], \tag{28}$$

$$[e, \Omega]\frac{d\Omega}{dt} = \frac{na^2 e \cos i}{\sqrt{1 - e^2}}\frac{d\Omega}{dt} = -L_1 \frac{\partial r_{12}}{\partial e} - (L_2 + \sum_{j=3}^{n} L_{3j}) \frac{\partial z_{12}}{\partial e}$$

$$+ \sum_{j=3}^{n} \left[L_{4j} \frac{\partial}{\partial e} H_j - L_{5j} \frac{\partial}{\partial e} \left(\frac{1}{r_{2j}} \right) \right], \tag{29}$$

$$[e, \omega] \frac{d\omega}{dt} = \frac{na^2 e}{\sqrt{1 - e^2}} \frac{d\omega}{dt} = - L_1 \frac{\partial r_{12}}{\partial e} - \left(L_2 + \sum_{j=3}^{n} L_{3j} \right) \frac{\partial z_{12}}{\partial e}$$

$$+ \sum_{j=3}^{n} \left[L_{4j} \frac{\partial}{\partial e} H_j - L_{5j} \frac{\partial}{\partial e} \left(\frac{1}{r_{2j}} \right) \right], \tag{30}$$

$$[i, \Omega] \frac{d\Omega}{dt} = na^2 \frac{d\Omega}{dt} \sqrt{1 - e^2} \sin i = - L_1 \frac{\partial r_{12}}{\partial i} - \left(L_2 + \sum_{j=3}^{n} L_{3j} \right) \frac{\partial z_{12}}{\partial i}$$

$$+ \sum_{j=3}^{n} \left[L_{4j} \frac{\partial}{\partial i} H_j - L_{5j} \frac{\partial}{\partial i} \left(\frac{1}{r_{2j}} \right) \right], \tag{31}$$

$$[a, T] \frac{dT}{dt} = \frac{1}{2} n^2 a \frac{dT}{dt} = - L_1 \frac{\partial r_{12}}{\partial a} - \left(L_2 + \sum_{j=3}^{n} L_{3j} \right) \frac{\partial z_{12}}{\partial a}$$

$$+ \sum_{j=3}^{n} \left[L_{4j} \frac{\partial}{\partial a} H_j - L_{5j} \frac{\partial}{\partial a} \left(\frac{1}{r_{2j}} \right) \right]; \tag{32}$$

with their respective negative expressions

$$[a, \; \Omega] \; \frac{d\Omega}{dt} = -\frac{1}{2} \, na \, \frac{d\Omega}{dt} \, \cos i \, \sqrt{1 - e^2} = -L_1 \, \frac{\partial r_{12}}{\partial a} - \left(L_2 + \sum_{j=3}^{n} L_{3j}\right) \frac{\partial z_{12}}{\partial a}$$

$$+ \sum_{j=3}^{n} \left[L_{4j} \, \frac{\partial}{\partial a} \, H_j - L_{5j} \, \frac{\partial}{\partial a} \left(\frac{1}{r_{2j}}\right)\right], \tag{33}$$

$$[a, \; \omega] \; \frac{d\omega}{dt} = -\frac{1}{2} \, na \, \frac{d\omega}{dt} \, \sqrt{1 - e^2} = -L_1 \, \frac{\partial r_{12}}{\partial a} - \left(L_2 + \sum_{j=3}^{n} L_{3j}\right) \frac{\partial z_{12}}{\partial a}$$

$$+ \sum_{j=3}^{n} \left[L_{4j} \, \frac{\partial}{\partial a} \, H_j - L_{5j} \, \frac{\partial}{\partial a} \left(\frac{1}{r_{2j}}\right)\right], \tag{34}$$

$$[\Omega, \; e] \; \frac{de}{dt} = -\frac{na^2 e \cos i}{\sqrt{1 - e^2}} \, \frac{de}{dt} = -L_1 \, \frac{\partial r_{12}}{\partial \Omega} - \left(L_2 + \sum_{j=3}^{n} L_{3j}\right) \frac{\partial z_{12}}{\partial \Omega}$$

$$+ \sum_{j=3}^{n} \left[L_{4j} \, \frac{\partial}{\partial \Omega} \, H_j - L_{5j} \, \frac{\partial}{\partial \Omega} \left(\frac{1}{r_{2j}}\right)\right], \tag{35}$$

$$[\omega, e] \frac{de}{dt} = -\frac{na^2 e}{\sqrt{1 - e^2}} \frac{de}{dt} = -L_1 \frac{\partial r_{12}}{\partial \omega} - \left(L_2 + \sum_{j=3}^{n} L_{3j}\right) \frac{\partial z_{12}}{\partial \omega}$$

$$+ \sum_{j=3}^{n} \left[L_{4j} \frac{\partial}{\partial \omega} H_j - L_{5j} \frac{\partial}{\partial \omega} \left(\frac{1}{r_{2j}}\right)\right], \tag{36}$$

$$[\Omega, i] \frac{di}{dt} = -na^2 \frac{di}{dt} \sqrt{1 - e^2} \sin i = -L_1 \frac{\partial r_{12}}{\partial \Omega} - \left(L_2 + \sum_{j=3}^{n} L_{3j}\right) \frac{\partial z_{12}}{\partial \Omega}$$

$$+ \sum_{j=3}^{n} \left[L_{4j} \frac{\partial}{\partial \Omega} H_j - L_{5j} \frac{\partial}{\partial \Omega} \left(\frac{1}{r_{2j}}\right)\right], \tag{37}$$

$$[T, a] \frac{da}{dt} = -\frac{1}{2} n^2 a \frac{da}{dt} = -L_1 \frac{\partial r_{12}}{\partial T} - \left(L_2 + \sum_{j=3}^{n} L_{3j}\right) \frac{\partial z_{12}}{\partial T}$$

$$+ \sum_{j=3}^{n} \left[L_{4j} \frac{\partial}{\partial T} H_j - L_{5j} \frac{\partial}{\partial T} \left(\frac{1}{r_{2j}}\right)\right]. \tag{38}$$

Matching terms of equal $\partial/\partial \alpha_1$ yields

$$\frac{1}{2} na \frac{da}{dt} \cos i \sqrt{1 - e^2} - \frac{na^2 e \cos i}{\sqrt{1 - e^2}} \frac{de}{dt} - na^2 \frac{di}{dt} \sqrt{1 - e^2} \sin i$$

$$- L_1 \frac{\partial r_{12}}{\partial \Omega} - \left(L_2 + \sum_{j=3}^{n} L_{3j}\right) \frac{\partial z_{12}}{\partial \Omega} + \sum_{j=3}^{n} \left[L_{4j} \frac{\partial}{\partial \Omega} H_j \right.$$

$$\left. - L_{5j} \frac{\partial}{\partial \Omega} \left(\frac{1}{r_{2j}}\right)\right], \quad \tag{39}$$

$$\frac{1}{2} n^2 a \frac{dT}{dt} - \frac{1}{2} na \frac{d\Omega}{dt} \cos i \sqrt{1 - e^2} - \frac{1}{2} na \frac{d\omega}{dt} \sqrt{1 - e^2} =$$

$$- L_1 \frac{\partial r_{12}}{\partial a} - \left(L_2 + \sum_{j=3}^{n} L_{3j}\right) \frac{\partial z_{12}}{\partial a} + \sum_{j=3}^{n} \left[L_{4j} \frac{\partial}{\partial a} H_j - L_{5j} \frac{\partial}{\partial a} \left(\frac{1}{r_{2j}}\right)\right], \tag{40}$$

$$\frac{1}{2} na \frac{da}{dt} \sqrt{1 - e^2} - \frac{na^2 e}{\sqrt{1 - e^2}} \frac{de}{dt} = - L_1 \frac{\partial r_{12}}{\partial \omega} - \left(L_2 + \sum_{j=3}^{n} L_{3j}\right) \frac{\partial z_{12}}{\partial \omega}$$

$$+ \sum_{j=3}^{n} \left[L_{4j} \frac{\partial}{\partial \omega} H_j - L_{5j} \frac{\partial}{\partial \omega} \left(\frac{1}{r_{2j}}\right)\right], \tag{41}$$

$$\frac{na^2 e \cos i}{\sqrt{1 - e^2}} \frac{d\Omega}{dt} + \frac{na^2 e}{\sqrt{1 - e^2}} \frac{d\omega}{dt} = -L_1 \frac{\partial r_{12}}{\partial e} - \left(L_2 + \sum_{j=3}^{n} L_{3j}\right) \frac{\partial z_{12}}{\partial e}$$

$$+ \sum_{j=3}^{n} \left[L_{4j} \frac{\partial}{\partial e} H_j - L_{5j} \frac{\partial}{\partial e} \left(\frac{1}{r_{2j}}\right)\right], \qquad (42)$$

$$-\frac{1}{2} n^2 a \frac{da}{dt} = -L_1 \frac{\partial r_{12}}{\partial T} - \left(L_2 + \sum_{j=3}^{n} L_{3j}\right) \frac{\partial z_{12}}{\partial T}$$

$$+ \sum_{j=3}^{n} \left[L_{4j} \frac{\partial}{\partial T} H_j - L_{5j} \frac{\partial}{\partial T} \left(\frac{1}{r_{2j}}\right)\right], \qquad (43)$$

$$na^2 \frac{d\Omega}{dt} \sqrt{1 - e^2} \sin i = -L_1 \frac{\partial r_{12}}{\partial i} - \left(L_2 + \sum_{j=3}^{n} L_{3j}\right) \frac{\partial z_{12}}{\partial i}$$

$$+ \sum_{j=3}^{n} \left[L_{4j} \frac{\partial}{\partial i} H_j - L_{5j} \frac{\partial}{\partial i} \left(\frac{1}{r_{2j}}\right)\right] \qquad (44)$$

Once the matching and recollection of terms in $\partial/\partial \alpha_l$ is done, the next obvious step is to solve for the unknown time rates of change of the elements;

i.e., da/dt, de/dt, di/dt, dω/dt, dΩ/dt and dT/dt. These are found by treating (39) through (44) simultaneously, since the number of unknowns equals the number of equations.

After a few simple algebraic manipulations it can be seen that

$$\frac{da}{dt} = -\frac{2}{n^2 a} \left\{ -L_1 \frac{\partial r_{12}}{\partial T} - \left(L_2 + \sum_{j=3}^{n} L_{3j}\right) \frac{\partial z_{12}}{\partial T}\right.$$

$$\left. + \sum_{j=3}^{n} \left[L_{4j} \frac{\partial}{\partial T} H_j - L_{5j} \frac{\partial}{\partial T} \left(\frac{1}{r_{2j}}\right)\right]\right\}, \qquad (45a)$$

$$\frac{de}{dt} = -\frac{(1-e^2)}{n^2 a^2 e} \left\{ -L_1 \frac{\partial r_{12}}{\partial T} - \left(L_2 + \sum_{j=3}^{n} L_{3j}\right) \frac{\partial z_{12}}{\partial T} + \sum_{j=3}^{n} \left[L_{4j} \frac{\partial}{\partial T} H_j \right.\right.$$

$$\left.\left. - L_{5j} \frac{\partial}{\partial T} \left(\frac{1}{r_{2j}}\right)\right]\right\} - \frac{\sqrt{1-e^2}}{n a^2 e} \left\{ -L_1 \frac{\partial r_{12}}{\partial \omega} - \left(L_2 + \sum_{j=3}^{n} L_{3j}\right) \frac{\partial z_{12}}{\partial \omega}\right.$$

$$\left. + \sum_{j=3}^{n} \left[L_{4j} \frac{\partial}{\partial \omega} H_j - L_{5j} \frac{\partial}{\partial \omega} \left(\frac{1}{r_{2j}}\right)\right]\right\}, \qquad (45b)$$

$$\frac{di}{dt} = \frac{\cos i}{na^2 \sin i \sqrt{1 - e^2}} \left\{ - L_1 \frac{\partial r_{12}}{\partial \omega} - \left(L_2 + \sum_{j=3}^{n} L_{3j} \right) \frac{\partial z_{12}}{\partial \omega} \right.$$

$$+ \sum_{j=3}^{n} \left[L_{4j} \frac{\partial}{\partial \omega} H_j - L_{5j} \frac{\partial}{\partial \omega} \left(\frac{1}{r_{2j}} \right) \right] \right\} - \frac{1}{na^2 \sin i \sqrt{1 - e^2}} \left\{ - L_1 \frac{\partial r_{12}}{\partial \Omega} \right.$$

$$\left. - \left(L_2 + \sum_{j=3}^{n} L_{3j} \right) \frac{\partial z_{12}}{\partial \Omega} + \sum_{j=3}^{n} \left[L_{4j} \frac{\partial}{\partial \Omega} H_j - L_{5j} \frac{\partial}{\partial \Omega} \left(\frac{1}{r_{2j}} \right) \right] \right\}, \qquad (45c)$$

$$\frac{d\omega}{dt} = \frac{\sqrt{1 - e^2}}{na^2 e} \left\{ - L_1 \frac{\partial r_{12}}{\partial e} - \left(L_2 + \sum_{j=3}^{n} L_{3j} \right) \frac{\partial z_{12}}{\partial e} + \sum_{j=3}^{n} \left[L_{4j} \frac{\partial}{\partial e} H_j \right. \right.$$

$$\left. - L_{5j} \frac{\partial}{\partial e} \left(\frac{1}{r_{2j}} \right) \right] \right\} - \frac{\cos i}{na^2 \sin i \sqrt{1 - e^2}} \left\{ - L_1 \frac{\partial r_{12}}{\partial i} - \left(L_2 + \sum_{j=3}^{n} L_{3j} \right) \frac{\partial z_{12}}{\partial i} \right.$$

$$+ \sum_{j=3}^{n} \left[L_{4j} \frac{\partial}{\partial i} H_j - L_{5j} \frac{\partial}{\partial i} \left(\frac{1}{r_{2j}} \right) \right] \right\}, \qquad (45d)$$

$$\frac{d\Omega}{dt} = \frac{1}{na^2 \sin i \sqrt{1 - e^2}} \left\{ - L_1 \frac{\partial r_{12}}{\partial i} - \left(L_2 + \sum_{j=3}^{n} L_{3j} \right) \frac{\partial z_{12}}{\partial i} \right.$$

$$+ \sum_{j=3}^{n} \left[L_{4j} \frac{\partial}{\partial i} H_j - L_{5j} \frac{\partial}{\partial i} \left(\frac{1}{r_{2j}}\right)\right]\}, \qquad (45e)$$

$$\frac{dT}{dt} = \frac{2}{n^2 a} \left\{ - L_1 \frac{\partial r_{12}}{\partial a} - \left(L_2 + \sum_{j=3}^{n} L_{3j}\right) \frac{\partial z_{12}}{\partial a} \right.$$

$$+ \sum_{j=3}^{n} \left[L_{4j} \frac{\partial}{\partial a} H_j - L_{5j} \frac{\partial}{\partial a} \left(\frac{1}{r_{2j}}\right)\right]\} + \frac{(1 - e^2)}{n^2 a^2 e} \left\{ - L_1 \frac{\partial r_{12}}{\partial e} \right.$$

$$- \left(L_2 + \sum_{j=3}^{n} L_{3j}\right) \frac{\partial z_{12}}{\partial e} + \sum_{j=3}^{n} \left[L_{4j} \frac{\partial}{\partial e} H_j - L_{5j} \frac{\partial}{\partial e} \left(\frac{1}{r_{2j}}\right)\right]\}. \qquad (45f)$$

E. Contributions of $\partial r_{12}/\partial \alpha_1$, $\partial z_{12}/\partial \alpha_1$, $\partial H_j/\partial \alpha_1$, $\partial (1/r_{2j})/\partial \alpha_1$ Involved in the Time Rates of Change of the Orbital Elements of m_2

How can one compute the $d\alpha_K/dt$ from the above six equations if the variability of r_{12}, z_{12}, H_j and r_{2j}^{-1} with respect to the α_1's is still unknown? To answer this question, with the help of Fig. 4, let p_{12} be a unit vector in the direction of perihelion, and let $\hat{h}_{12}$ be a unit vector on the orbital plane of m_2 and orthogonal to p_{12}. Then

$$\hat{P}_{12} = (\cos \omega \cos \Omega - \sin \omega \sin \Omega \cos i) \ \hat{x}$$

$$+ (\cos \omega \sin \Omega + \sin \omega \cos \Omega \cos i) \ \hat{y} + \sin \omega \sin i \ \hat{z}, \qquad (46a)$$

$$\hat{h}_{12} = (-\sin \omega \cos \Omega - \cos \omega \sin \Omega \cos i) \ \hat{x}$$

$$+ (\cos \omega \cos \Omega \cos i - \sin \omega \sin \Omega) \ \hat{y} + \cos \omega \sin i \ \hat{z}, \qquad (46b)$$

for the orbit described by m_2; and

$$\hat{P}_j = (\cos \omega_j \cos \Omega_j - \sin \omega_j \sin \Omega_j \cos i_j) \ \hat{x}$$

$$+ (\cos \omega_j \sin \Omega_j + \sin \omega_j \cos \Omega_j \cos i_j) \ \hat{y} + \sin \omega_j \sin i_j \ \hat{z}, \qquad (46c)$$

$$\hat{h}_j = (-\sin \omega_j \cos \Omega_j - \cos \omega_j \sin \Omega_j \cos i_j) \ \hat{x}$$

$$+ (\cos \omega_j \cos \Omega_j \cos i_j - \sin \omega_j \sin \Omega_j) \ \hat{y} + \cos \omega_j \sin i_j \ \hat{z}, \qquad (46d)$$

for the orbit described by the planet whose mass is given by m_j, and whose elements are given by a_j, e_j, i_j, ω_j, and Ω_j; with $j = 3, 4, \ldots, n$.

Using these unit vectors $\hat{P}_{12}$, $\hat{h}_{12}$, $\hat{P}_j$ and $\hat{h}_j$ of (46), and comparing with Fig. 4, the radial vectors $\vec{r}_{12}$ and $\vec{r}_{1j}$ are represented by

$$\vec{r}_{12} = p_{12}\hat{p}_{12} + h_{12}\hat{h}_{12},$$

$$\vec{r}_{1j} = p_j\hat{p}_j + h_j\hat{j}_j. \tag{47}$$

Letting E be the eccentric anomaly of an ellipse, from Fig. 5 one sees that

$$\cos E = e + \left(\frac{r_{12}}{a}\right) \cos \phi, \tag{48a}$$

$$\sin E = \frac{p_2 p_3}{a} + \left(\frac{r_{12}}{a}\right) \sin \phi, \text{ and} \tag{48b}$$

$$(1 - e^2) = \left[\frac{p_1 p_2}{p_1 p_3}\right]^2. \tag{48c}$$

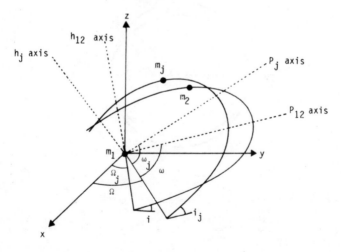

Fig. 4. Position of the p_{12}, h_{12}, p_j and h_j axes on the orbits described by m_2 and m_j respectively.

Since $p_{12} = r_{12} \cos \phi$ and $h_{12} = r_{12} \sin \phi$, it is obvious from (48) that

$$p_{12} = a(\cos E - e); \quad h_{12} = a \sin E \sqrt{1 - e^2}, \quad (49)$$

$$p_j = a_j(\cos E_j - e_j); \quad h_j = a_j \sin E_j \sqrt{1 - e_j^2},$$

where E and E_j are the eccentric anomalies in the orbits of m_2 and m_j, in that order.

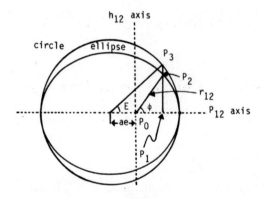

Fig. 5.   Illustration of the angle E defined as the eccentric anomaly. The angle ϕ lies between the perihelion axis and the orbital radius r_{12}. P_o is the focus at which the sun is located. Observe the orthogonality among the p_{12} and h_{12} axes.

Furthermore, since the evaluation of the orbits is taken when $\vec{r}_{12}$, $\vec{r}_{1j}$, etc., are oriented towards both aphelion and perihelion*, $E = E_j$ equals π and zero respectively, thus reducing (49) to[+]

$$p_{12} = \mp a (1 \pm e),$$

$$p_j = \mp a_j (1 \pm e_j), \qquad (50)$$

$$h_{12} = h_j = 0.$$

Equations (47) and (50) imply that

$$r_{12} = a (1 \pm e), \quad r_{1j} = a_j (1 \pm e_j). \qquad (51)$$

* Evaluation at both aphelion and perihelion allows to reduce the equations notably and isolates the behavior of $d\omega/dt$ on m_2 due to m_1 and Σ_{mj}.

[+] In the equations to come the upper signs will apply to the aphelion evaluation and the lower signs will apply to the perihelion evaluation.

Consequently, the orbital radii do not involve the elements i, ω, Ω and T. Therefore, at aphelion and perihelion

$$\frac{\partial r_{12}}{\partial a} = (1 \pm e), \tag{52a}$$

$$\frac{\partial r_{12}}{\partial e} = \pm a, \tag{52b}$$

$$\frac{\partial r_{12}}{\partial i, \omega, \Omega, T} = 0. \tag{52c}$$

Differentiation of z_{12} with respect to the a_1's is found by recalling from Fig. 4, (46a), (47) and (51) that at both aphelion and perihelion:

$$z_{12} = a (1 \pm e) \sin \omega \sin i. \tag{53}$$

Thus

$$\frac{\partial z_{12}}{\partial a} = (1 \pm e) \sin \omega \sin i, \tag{54a}$$

$$\frac{\partial z_{12}}{\partial e} = \pm a \sin \omega \sin i, \tag{54b}$$

$$\frac{\partial z_{12}}{\partial i} = a (1 \pm e) \sin \omega \cos i, \tag{54c}$$

$$\frac{\partial z_{12}}{\partial \omega} = a \ (1 \pm e) \ \cos \omega \sin i \qquad (54d)$$

$$\frac{\partial z_{12}}{\partial \Omega, T} = 0. \qquad (54e)$$

Before finding the contribution of $\partial H_j / \partial \alpha_1$ for equations (45), one must know how r_{2j}^{-1} and $\vec{r}_{12} \cdot \vec{r}_{1j}/r_{1j}^{3}$ vary with respect to the orbital elements.

By recalling (10a), one has

$$\frac{\partial}{\partial \alpha_1} \left(\frac{1}{r_{2j}}\right) = \left(\frac{1}{r_{2j}^{3}}\right) \ [(x_{1j} - x_{12}) \ \frac{\partial x_{12}}{\partial \alpha_1} + \qquad (55a)$$

$$(y_{1j} - y_{12}) \ \frac{\partial y_{12}}{\partial \alpha_1} + (z_{1j} - z_{12}) \ \frac{\partial z_{12}}{\partial \alpha_1}],$$

where at both aphelion and perihelion, and from Fig. 4,

$$x_{12} = a(1 \pm e) \ [\cos \omega \cos \Omega - \sin \omega \sin \Omega \cos i], \qquad (55b)$$

$$x_{1j} = a_j(1 \pm e_j) \ [\cos \omega_j \cos \Omega_j - \sin \omega_j \sin \Omega_j \cos i_j],$$

$$y_{12} = a(1 \pm e) \ [\cos \omega \sin \Omega + \sin \omega \cos \Omega \cos i], \qquad (55c)$$

$$y_{1j} = a_j(1 \pm e_j) \left[\cos \omega_j \sin \Omega_j + \sin \omega_j \cos \Omega_j \cos i_j\right],$$

$$z_{12} = a(1 \pm e) \sin \omega \sin i, \qquad (55d)$$

$$z_{1j} = a_j(1 \pm e_j) \sin \omega_j \sin i_j.$$

After performing the differentiation of x_{12}, y_{12} and z_{12}, with respect to the α_1's, and placing these into (55a) yields, for each of the six orbital elements, the following expressions.

$$\frac{\partial}{\partial a} \left(\frac{1}{r_{2j}}\right) = \frac{1}{r_{2j}^3} \{ [a_j(1 \pm e_j) (\cos \omega_j \cos \Omega_j - \sin \omega_j \sin \Omega_j \cos i_j) - $$

$$a(1 \pm e) (\cos \omega \cos \Omega - \sin \omega \sin \Omega \cos i)] \left[(1 \pm e) (\cos \omega \cos \Omega - \right.$$

$$\sin \omega \sin \Omega \cos i] + [a_j(1 \pm e_j) (\cos \omega_j \sin \Omega_j + \sin \omega_j \cos \Omega_j \cos i_j) - $$

$$a(1 \pm e) (\cos \omega \sin \Omega + \sin \omega \cos \Omega \cos i)] \left[(1 \pm e) \cos \omega \sin \Omega + \right.$$

$$\sin \omega \cos \Omega \cos i] + [a_j(1 \pm e_j) \sin \omega_j \sin i_j - a(1 \pm e) \sin \omega \sin i]$$

$$[(1 \pm e) \sin \omega \sin i]\}, \qquad (56a)$$

$$\frac{\partial}{\partial e} \left(\frac{1}{r_{2j}}\right) = \frac{1}{r_{2j}^3} \{ [a_j(1 \pm e_j) (\cos \omega_j \cos \Omega_j - \sin \omega_j \sin \Omega_j \cos i_j) - $$

a(1 ± e) (cos ω cos Ω - sin ω sin Ω cos i)] [± a(cos ω cos Ω -

sin ω sin Ω cos i)] + [a_j(1 ± e_j) (cos ω_j sin Ω_j + sin ω_j cos Ω_j cos i_j)

- a(1 ± e) (cos ω sin Ω + sin ω cos Ω cos i)] [± a(cos ω sin Ω +

sin ω cos Ω cos i)] + [a_j(1 ± e_j) sin ω_j sin i_j - a(1 ± e) sin ω sin i]

[± a sin ω sin i]},   (56b)

$$\frac{\partial}{\partial i} \left(\frac{1}{r_{2j}}\right) = \frac{1}{r_{2j}^3} \{[a_j(1 \pm e_j) (\cos \omega_j \cos \Omega_j - \sin \omega_j \sin \Omega_j \cos i_j)$$

- a(1 ± e) (cos ω cos Ω - sin ω sin Ω cos i)] [a(1 ± e) sin ω sin Ω sin i]

+ [a_j(1 ± e_j) (cos ω_j sin Ω_j + sin ω_j cos Ω_j cos i_j) - a(1 ± e) (cos ω

sin Ω + sin ω cos Ω cos i] [- a(1 ± e) sin ω cos Ω sin i]

+ [a_j(1 ± e_j) sin ω_j sin i_j - a(1 ± e) sin ω sin i] [a(1 ± e) sin ω

cos i]},   (56c)

$$\frac{\partial}{\partial \omega} \left(\frac{1}{r_{2j}}\right) = \frac{1}{r_{2j}^3} \{[a_j(1 \pm e_j) (\cos \omega_j \cos \Omega_j - \sin \omega_j \sin \Omega_j \cos i_j)$$

- a(1 ± e) (cos ω cos Ω - sin ω sin Ω cos i)] [a(1 ± e) (-sin ω cos Ω

$- \cos \omega \sin \Omega \cos i)] + [a_j(1 \pm e_j) (\cos \omega_j \sin \Omega_j + \sin \omega_j \cos \Omega_j \cos i_j)$

$- a(1 \pm e) (\cos \omega \sin \Omega + \sin \omega \cos \Omega \cos i)] [a(1 \pm e) (-\sin \omega \sin \Omega +$

$\cos \omega \cos \Omega \cos i)] + [a_j(1 \pm e_j) \sin \omega_j \sin i_j - a(1 \pm e) \sin \omega \sin i]$

$$[a(1 \pm e) \cos \omega \sin i]\}, \tag{56d}$$

$$\frac{\partial}{\partial \Omega} \left(\frac{1}{r_{2j}}\right) = \frac{1}{r_{2j}^3} \{[a_j(1 \pm e_j) (\cos \omega_j \cos \Omega_j - \sin \omega_j \sin \Omega_j \cos i_j)$$

$- a(1 \pm e) (\cos \omega \cos \Omega - \sin \omega \sin \Omega \cos i)] [a(1 \pm e) (-\sin \Omega \cos \omega$

$- \cos \Omega \sin \omega \cos i)] + [a_j(1 \pm e_j) (\cos \omega_j \sin \Omega_j + \sin \omega_j \cos \Omega_j \cos i_j)$

$- a(1 \pm e) (\cos \omega \sin \Omega + \sin \omega \cos \Omega \cos i)] [a(1 \pm e) (\cos \omega \cos \Omega$

$$- \sin \omega \sin \Omega \cos i]\}, \tag{56e}$$

$$\frac{\partial}{\partial T} \left(\frac{1}{r_{2j}}\right) = 0. \tag{56f}$$

In expressions (56), $r_{2j}{}^3$ is given by

$r_{2j}{}^3 = \{[a_j(1 \pm e_j) (\cos \omega_j \cos \Omega_j - \sin \omega_j \sin \Omega_j \cos i_j) - a(1 \pm e)$

$(\cos \omega \cos \Omega - \sin \omega \sin \Omega \cos i)]^2 + [a_j(1 \pm e_j) (\cos \omega_j \sin \Omega_j$

$$+ \sin \omega_j \cos \Omega_j \cos i_j) - a(1 \pm e)(\cos \omega \sin \Omega + \sin \omega \cos \Omega \cos i)]^2$$

$$+ [a_j(1 \pm e_j) \sin \omega_j \sin i_j - a(1 \pm e) \sin \omega \sin i]^2\}^{3/2}. \qquad (57)$$

From (10d), the scalar product of $\vec{r}_{12} \cdot \vec{r}_{1j}/r_{1j}{}^3$ can also be expressed as partial differentiations with respect to the elements α_1's. Thus

$$\frac{\partial}{\partial \alpha_1} \left(\frac{\vec{r}_{12} \cdot \vec{r}_{1j}}{r_{1j}{}^3} \right) = \frac{1}{r_{1j}{}^3} \left[x_{1j} \frac{\partial x_{12}}{\partial \alpha_1} + y_{1j} \frac{\partial y_{12}}{\partial \alpha_1} + z_{1j} \frac{\partial z_{12}}{\partial \alpha_1} \right], \qquad (58)$$

which, upon using equations (56) with their correspondent differentiations relative to a, e, i, ω, Ω and T individually, yield

$$\frac{\partial}{\partial a} \left(\frac{\vec{r}_{12} \cdot \vec{r}_{1j}}{r_{1j}{}^3} \right) = \frac{(1 \pm e)}{a_j{}^2(1 \pm e_j)^2} [(\cos \omega_j \cos \Omega_j - \sin \omega_j \sin \Omega_j \cos i_j)$$

$$(\cos \omega \cos \Omega - \sin \omega \sin \Omega \cos i) + (\cos \omega_j \sin \Omega_j + \sin \omega_j \cos \Omega_j \cos i_j)$$

$$(\cos \omega \sin \Omega + \sin \omega \cos \Omega \cos i) + \sin \omega_j \sin i_j \sin \omega \sin i], \qquad (59a)$$

$$\frac{\partial}{\partial e} \left(\frac{\vec{r}_{12} \cdot \vec{r}_{1j}}{r_{1j}{}^3} \right) = \frac{\pm a}{a_j{}^2(1 \pm e_j)^2} [(\cos \omega_j \cos \Omega_j - \sin \omega_j \sin \Omega_j \cos i_j)$$

$(\cos \omega \cos \Omega - \sin \omega \sin \Omega \cos i) + (\cos \omega_j \sin \Omega_j + \sin \omega_j \cos \Omega_j \cos i_j)$

$(\cos \omega \sin \Omega + \sin \omega \cos \Omega \cos i) + \sin \omega_j \sin i_j \sin \omega \sin i$ \hfill (59b)

$$\frac{\partial}{\partial i} \left(\frac{\vec{r}_{12} \cdot \vec{r}_{1j}}{r_{1j}^3} \right) = \frac{a(1 \pm e)}{a_j^2 (1 \pm e_j)^2} \left[(\cos \omega_j \cos \Omega_j - \sin \omega_j \sin \Omega_j \cos i_j) \right.$$

$(\sin \omega \sin \Omega \sin i) - (\cos \omega_j \sin \Omega_j + \sin \omega_j \cos \Omega_j \cos i_j)$

$(\sin \omega \cos \Omega \sin i) + \sin \omega_j \sin i_j \sin \omega \cos i],$ \hfill (59c)

$$\frac{\partial}{\partial \omega} \left(\frac{\vec{r}_{12} \cdot \vec{r}_{1j}}{r_{1j}^3} \right) = \frac{a(1 \pm e)}{a_j^2 (1 \pm e_j)^2} \left[(\cos \omega_j \cos \Omega_j - \sin \omega_j \sin \Omega_j \cos i_j) \right.$$

$(- \sin \omega \cos \Omega - \cos \omega \sin \Omega \cos i) + (\cos \omega_j \sin \Omega_j + \sin \omega_j \cos \Omega_j \cos i_j)$

$(- \sin \omega \sin \Omega + \cos \omega \cos \Omega \cos i) + \sin \omega_j \sin i_j \cos \omega \sin i],$ \hfill (59d)

$$\frac{\partial}{\partial \Omega} \left(\frac{\vec{r}_{12} \cdot \vec{r}_{1j}}{r_{1j}^3} \right) = \frac{a(1 \pm e)}{a_j^2 (1 \pm e_j)^2} \left[(\cos \omega_j \cos \Omega_j - \sin \omega_j \sin \Omega_j \cos i_j) \right.$$

$(- \sin \Omega \cos \omega - \cos \Omega \sin \omega \cos i) + (\cos \omega_j \sin \Omega_j + \sin \omega_j \cos \Omega_j \cos i_j)$

$$(\cos \omega \cos \Omega - \sin \omega \sin \Omega \cos i)\,], \qquad (59e)$$

$$\frac{\partial}{\partial T} \left(\frac{\vec{r}_{12} \cdot \vec{r}_{1j}}{r_{1j}^3} \right) = 0. \qquad (59f)$$

By making the proper combinations of (56) and (59) according to the perturbative function H_j of (10d) yields the contributions of $\partial H_j/\partial \alpha_1$. Therefore, in general terms

$$\frac{\partial H_j}{\partial \alpha_1} = \frac{\partial}{\partial \alpha_1} \left(\frac{1}{r_{2j}} \right) - \frac{\partial}{\partial \alpha_1} \left(\frac{\vec{r}_{12} \cdot \vec{r}_{1j}}{r_{1j}^3} \right),$$

and for each of the orbital elements

$$\frac{\partial H_j}{\partial a} = (56a) - (59a), \qquad (60a)$$

$$\frac{\partial H_j}{\partial e} = (56b) - (59b), \qquad (60b)$$

$$\frac{\partial H_j}{\partial i} = (56c) - (59c), \qquad (60c)$$

$$\frac{\partial H_j}{\partial \omega} = (56d) - (59d), \qquad (60d)$$

$$\frac{\partial H_j}{\partial \Omega} = (56e) - (59e), \qquad\qquad (60e)$$

$$\frac{\partial H_j}{\partial T} = (56f) - (59f), \qquad\qquad (60f)$$

Once the partial differentiations of r_{12}, z_{12}, H_j and r_{2j}^{-1} relative to the elements are found, when the evaluation takes place at perihelion and aphelion, the only remaining step of the theory is to perform substitutions of (52), (54), (56) and (60) into the expressions for da/dt, de/dt, di/dt, dω/dt, dΩ/dt and dT/dt as given by (45).

In brief manipulation, it can be proved that the following equations are true.

$$\frac{da}{dt} = 0, \qquad\qquad (61a)$$

$$\frac{de}{dt} = \pm \frac{\sqrt{1 - e^2}}{na^2 e} \{- L_1 \ (52c) - (L_2 + \sum_{j=3}^{n} L_{3j}) \ (54d)$$

$$+ \sum_{j=3}^{n} [L_{4j} \ (60d) - L_{5j} \ (56d)]\}, \qquad\qquad (61b)$$

$$\frac{di}{dt} = \frac{\mp \cos i}{na^2 \sin i \sqrt{1 - e^2}} \left\{ - L_1 \text{ (52c)} - (L_2 + \sum_{j=3}^{n} L_{3j}) \text{ (54d)} \right.$$

$$+ \sum_{j=3}^{n} \left[L_{4j} \text{ (60d)} - L_{5j} \text{ (56d)} \right] \right\} \pm \frac{1}{na^2 \sin i \sqrt{1 - e^2}} \left\{ - L_1 \text{ (52c)} \right.$$

$$- (L_2 + \sum_{j=3}^{n} L_{3j}) \text{ (54e)} + \sum_{j=3}^{n} \left[L_{4j} \text{ (60e)} - L_{5j} \text{ (56e)} \right] \right\}, \qquad (61c)$$

$$\frac{d\omega}{dt} \mp \frac{\sqrt{1 - e^2}}{na^2 e} \left\{ - L_1 \text{ (52b)} - (L_2 + \sum_{j=3}^{n} L_{3j}) \text{ (54b)} + \sum_{j=3}^{n} \left[L_{4j} \text{ (60b)} \right. \right.$$

$$- L_{5j} \text{ (56b)} \right] \right\} \pm \frac{\cos i}{na^2 \sin i \sqrt{1 - e^2}} \left\{ - L_1 \text{ (52c)} - (L_2 + \sum_{j=3}^{n} L_{3j}) \text{ (54c)} \right.$$

$$+ \sum_{j=3}^{n} \left[L_{4j} \text{ (60c)} - L_{5j} \text{ (56c)} \right] \right\}, \qquad (61d)$$

$$\frac{d\Omega}{dt} = \frac{\mp 1}{na^2 \sin i \sqrt{1 - e^2}} \left\{ - L_1 \text{ (52c)} - (L_2 + \sum_{j=3}^{n} L_{3j}) \text{ (54c)} \right.$$

$$+ \sum_{j=3}^{n} \left[L_{4j}(60c) - L_{5j}(56c) \right] \}, \tag{61e}$$

$$\frac{dT}{dt} = \frac{\mp 2}{n^2 a} \left\{ - L_1 (52a) - \left(L_2 + \sum_{j=3}^{n} L_{3j} \right) (54a) + \sum_{j=3}^{n} \left[L_{4j} (60a) \right. \right.$$

$$\left. - L_{5j} (56a) \right] \} \mp \frac{(1 - e^2)}{n^2 a^2 e} \left\{ - L_1 (52b) - \left(L_2 + \sum_{j=3}^{n} L_{3j} \right) (54b) \right.$$

$$+ \sum_{j=3}^{n} \left[L_{4j} (60b) - L_{5j} (56b) \right] \}. \tag{61f}$$

As one can see from the above six equations, the precession of the orbit described by m_2 around m_1 is given by the time rate of change of the argument of the perihelion, as expressed by (61d).

The motion in the line of nodes of m_2 is computed by means of (61e), which measures the time rate of change in the longitude of the ascending node. Observe that any change in the eccentricity of m_2 is found from (61b), and that there is no theoretical variation in the semi-major axis, when the evaluation of the $d\alpha_K/dt$ is taken precisely at both perihelion and aphelion.

Thus, by knowing all six elements for each of the nine planets of which the solar system is presumably composed, and by letting m_2 be Mercury, m_3 be Venus, . . ., m_{10} be Pluto, the numerical outcome of (61) can be obtained. This, along with analysis and interpretation of the results is given in the paragraphs to follow.

VI.

RESULTS ON MERCURY'S ORBITAL PRECESSION

Accordingly, the expressions given by (61) predict the time rates of change for all six elements of Mercury's orbit.

For the case of perihelion passage, the evaluation of the Lagrange brackets is that given by (26). At aphelion the resulting equations are as given by (26), except that they must be multiplied by -1 since the radial vector $\vec{r}_{12}$ points in opposite directions at perihelion and at aphelion.

The data used for the orbital elements of the planets was obtained from Allen's Astrophysical Quantities book.[26] Appendix C gives the data for each of the nine planets, according to the information gathered by Allen for the year 1973. Two computer programs were employed for the computation of the results. One of these programs performed the calculations when the planets are located at their

---

[26] Allen, C. W. *Astrophysical Quantities* (London: Athlone Press, 1973), pp. 140, 141.

respective perihelia, while the other did the calculations when the planets pass through their respective aphelia. Adding these two contributions gives the total change per unit time in each of the six elements of Mercury's orbit. However, due to a lack of library facilities, observational data for only the dynamical precession of the perihelion (dω/dt) was obtained. Thus, the numerical predictions found for the change in time of the other five elements will not be discussed at this time. Appendix D lists the constants used in the calculations.

The results of dω/dt correspond to the solar oblateness values of $3.6 \pm 0.23 \times 10^{-5}$, $4.51 \pm 0.34 \times 10^{-5}$, and $5.0 \pm 0.7 \times 10^{-5}$, found by Poor[27] in 1905, Dicke[28] in 1966, and again by Dicke and Goldenberg[29] in 1967, respectively.

These results are given below. For the solar oblateness value $\Delta r/r$, of $3.6 \pm 0.23 \times 10^{-5}$,

dω/dt = 570.86 ± 1.48" arc/century.

For the solar oblateness value,

[27] See footnotes 5 and 6.

[28] See footnote 20, p. 419.

[29] See footnote 18.

$\Delta r/r$, of $4.51 \pm 0.34 \times 10^{-5}$,

$d\omega/dt = 576.76 \pm 2.95"$ arc/century.

And for the solar oblateness value, $\Delta r/r$, of $5.0 \pm 0.7 \times 10^{-5}$,

$d\omega/dt = 579.71 \pm 5.90"$ arc/century.

As a complementary calculation, the average of the three previous oblateness values is considered. Thus, for the mean solar oblateness of $4.37 \pm 1.27 \times 10^{-5}$,

$d\omega/dt = 575.29 \pm 8.87"$ arc/century.

VII.

DISCUSSION

The final task of this monograph is the analyzation and interpretation of the results found in Chapter 6.

Recall that for the mean of the three solar oblateness values, $\dot{\omega} =$ 575.29 ± 8.87" arc/century. For Dicke and Goldenberg's oblateness value, $\dot{\omega} =$ 579.71 ± 5.90" arc/century. For Dicke's 1966 value, $\dot{\omega} = 576.76 ± 2.95"$ arc/century. And for Poor's value, $\dot{\omega} = 570.86 ± 1.48"$ arc/century.

How well do these predictions of Mercury's perihelion precession agree with observation? The accepted value as calculated by Classical Perturbation Theory is 532" arc/century.[30] This value differs by 43" arc/century from the observed precession of 575" arc/century.[31]

About 65 years ago Einstein applied his theory of General

---

[30] Price, M. P., and Rush, W. F. "Nonrelativistic contribution to Mercury's perihelion precession." *American Journal of Physics*, Vol. 47, No. 6, June 1979, p. 534.

[31] *Ibid.*

Relativity to the problem of account-
ing for the 43" arc/century discrepancy. Coincidentally, or otherwise his
predicted value agreed exactly with
the unexplained anomalous motion of
Mercury's perhelion.

This monograph then was developed
in 1982 from pure Classical Mechanics
with the purpose of re-explaining such
oddity by assuming a non-central gravitational force. Using the best data
available, calculations for the precession were made to within the errors
specified by the various values of
solar oblateness. By comparing the
results found with the currently
observed rotational rate of Mercury's
orbit, it is concluded that the
approach used here is excellent.

VIII.

CONCLUSIONS

The implications of the work of this monograph are of great importance since it is shown that the Newtonian cosmology does explain the phenomena of motion of Mercury's perihelion. The Newtonian cosmology has been abandoned by most cosmologists for the totally different Einsteinian cosmology in spite of the brilliant work of the late Herbert Dingle and G. Burniston Brown in showing the fatal inconsistencies in both the physical and mathematical structure of relativity; the lack of real physical evidence for relativity as pointed out by Ives, Essen, and others; and the development of explanations of the basic nature of the atom and electromagnetic radiation by Thomas G. Barnes and his students on a classical basis. The modern cosmologists, in the main, have thrown over a picture of the universe in which bodies are moved along by "alien forces" and, in choosing a set of axioms not grounded in common experience since they lack real proof, made a picture of the cosmos that is a system of free bodies moving along courses which offer them the least resistance.

It has been man's experience in seeking knowledge that the cosmos follows in the main the laws we have observed it to follow in the small rather than to behave according to ideas that some hold it should follow. In contrast to this experience modern cosmologists have returned to the discredited practice of inventing arbitrary general practices, with no justification except that they seem "right," and a fitting phenomena to the requirements of ideas. For example, following one of Einstein's assumptions we have been presented with the "cosmological prin- ciple" which demands that on a large scale though not on a small (where it could be tested), every part of the cosmos must be exactly the same as any other. Much of this approach was expressed by Sir Arthur Eddington when he says "there is nothing in the whole system of laws of physics that cannot be deduced unambiguously from episte- mological considerations. An intelli- gence, unacquainted with our universe, but with the system of thought by which the human mind interprets to itself the content of its sensory experience, should be able to attain all the knowledge of physics that we have attained by physics." Relativity has been applied to the universe by making assumptions which have become dogmas.

In modern-day cosmology based on relativity if we look at modern state- ments, the universe is no longer the

observable world that for so long was
the object of astronomical study but a
hypothetical entity of which what we
can observe is an almost negligible
part.   The assertions many cosmologists make about it seem rarely sus-
ceptible of test since they would
require much time into the past and
are obviously beyond any practical
test.   It seems that there is little
control of speculation and preserving
of legitimate theory from idle fancy.

Dingle has aptly remarked "that
science is not based on an axiom but
on the comparison of theories with
observation." Modern cosmology based
on relativity seems nothing but a
fantasy of mathematicians who find it
agreeable that the world should be
made in this way. We should abandon
Aristotle's approach to study of the
cosmos, forsaking mathematical arti-
fice, and develop a cosmology based on
observations.

APPENDICES

APPENDIX A

Derivation of equation (2a).

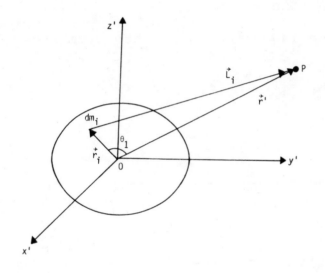

Fig. 1A Geometry for calculating the potential at a unit mass located at P due to an oblate mass distribution m_1.

Let

$$dV = - \frac{Gdm_i}{L_i}, \text{ where } \vec{L}_i = \vec{r} - \vec{r}_i, \quad (A1)$$

be the potential element produced by the mass element dm_i at P. If one introduces an angle θ_i between the directions of $\vec{r}_i$ and $\vec{r}$ and employs the law of cosines, then from Fig. 1A,

$$L_i = (r'^2 + r_i^2 - 2r'r_i \cos \theta_i)^{1/2} \quad (A2)$$

so that (A1) becomes

$$dV = \frac{-Gdm_i}{(r'^2 + r_i^2 - 2r'r_i \cos \theta_i)^{1/2}} \quad (A3)$$

If $r' > r_i$, the ratio r_i/r' is always less than unity. This means that one can take a power series expansion of this ratio.

Let $\quad \dfrac{1}{L_i} = \dfrac{1}{r'(1 + b)^{1/2}}$, $\qquad$ (A4)

where $b = -2 \left(\dfrac{r_i}{r'}\right) \cos \theta_i + \left(\dfrac{r_i}{r'}\right)^2$. $\quad$ (A5)

Using the binomial theorem of expansion,

$$(1+b)^{-n} = 1 - nb + \frac{n(n+1) b^2}{2!} - \frac{n(n+1)(n+2) b^3}{3!} + \ldots \qquad \text{(A6)}$$

and neglecting orders larger than $(r_i/r')^2$, with $n = \frac{1}{2}$, (A6) becomes,

$$(1+b)^{-\frac{1}{2}} \approx 1 - \frac{1}{2} \left[-2 \left(\frac{r_i}{r'} \right) \cos \theta_i + \left(\frac{r_i}{r'} \right)^2 \right] + \frac{3}{8} \left[-2 \left(\frac{r_i}{r'} \right) \cos \theta_i + \left(\frac{r_i}{r'} \right)^2 \right]^2$$

$$\approx 1 + \left(\frac{r_i}{r'} \right) \cos \theta_i + \left(\frac{r_i}{r'} \right)^2 \left(\frac{3}{2} \cos^2 \theta_i - \frac{1}{2} \right).$$

Hence multiplying this expression by $(r')^{-1}$ one finds that

$$V(\vec{r}') = - \frac{G}{r'} \int dm_i - \frac{G}{r'^2} \int r_i \cos \theta_i \, dm_i$$

$$- \frac{G}{2r'^3} \int r_i^2 \, dm_i \, (3 \cos^2 \theta_i - 1) + \ldots \qquad \text{(A7)}$$

(A7) can also be expressed as

$$V(\vec{r}') = V_M(\vec{r}') + V_D(\vec{r}') + V_Q(\vec{r}') + \ldots,$$

where M, D, and Q stand for monopole, dipole, and quadrupole terms of potential respectively.

The functions of θ_i which have arisen in (A7) are called Legendre polynomials. The integration of the first term of (A7) is the total mass of the spheroid. The second integral term is the dipole moment about the origin. Since the origin was chosen to lie at the center of the spheroid, the term is zero. The integral of the last term on the right-hand side of (A7) is identified as the quadrupole moment of the mass distribution.

If m_1 is the total mass of the spheroid, integration of the non-vanishing terms in expression (A7) yields

$$\frac{G}{r'} \int dm_i = \frac{Gm_1}{r'}, \qquad (A8)$$

$$\frac{G}{2r'^3} \left[3 \int r_i^2 \cos^2 \theta_i \, dm_i - \int r_i^2 \, dm_i \right] =$$

$$\frac{3G}{2r'^3} \int r_i^2 \cos^2 \theta_i \, dm_i - \frac{G}{2r'^3} \int r_i^2 \, dm_i =$$

$$\frac{3G}{2r'^3} \int r_i^2 (1 - \sin^2 \theta_i) \, dm_i - \frac{G}{2r'^3} \int r_i^2 \, dm_i =$$

$$\frac{G}{r'^3} \int r_i^2 \, dm_i - \frac{3G}{2r'^3} \int r_i^2 \sin^2 \theta_i \, dm_i, \qquad \text{(A9)}$$

where $\int r_i^2 dm_i = I_0$ (moment of inertia about the origin of m_1), and $\int r_i^2 \sin^2 \theta_i \, dm_i = I$ (moment of inertia about the line r').

    With (A8) and (A9) the potential at P is

$$V = -\frac{Gm_1}{r'} - \frac{GI_0}{r'^3} + \frac{3G}{2r'^3} I. \qquad \text{(A10)}$$

I_0 can be expressed in terms of m_1's principal moments of inertia $I_{x',y',z'}$ as

$$I_0 = \frac{1}{2} (I_{x'} + I_{y'} + I_{z'}). \qquad \text{(A11)}$$

Letting

$$l_{x'} = \frac{x'}{r'}, \quad l_{y'} = \frac{y'}{r'}, \quad l_{z'} = \frac{z'}{r'} \qquad \text{(A12)}$$

be the direction cosines of the line r', the moment of inertia I is written as

$$I = I_{x'} l_{x'}^2 + I_{y'} l_{y'}^2 + I_{z'} l_{z'}^2. \quad (A13)$$

From the x'y' symmetry of the spheriod,

$$I_{x'y'} = I_{x'} = I_{y'} \neq I_{z'}. \quad (A14)$$

Hence, (A11) and (A13) become

$$I_o = I_{x'y'} + I_{z'}/2, \quad (A15a)$$

$$I = I_{x'y'} (l_{x'}^2 + l_{y'}^2) + I_{z'} l_{z'}^2. \quad (A15b)$$

Using (A12) in (A15b) yields for I the following expression.

$$I = \frac{1}{r'^2} \left(I_{x'y'} (x'^2 + y'^2) + I_{z'} z'^2 \right).$$

So, $\quad I = I_{x'y'} - (I_{x'y'} - I_{z'}) \dfrac{z'^2}{r'^2}. \quad (A16)$

Substituting (A15a) and (A16) in the potential expression (A10) yields

$$V = -\frac{Gm_1}{r'} - \frac{G(I_{x'y'} + I_{z'}/2)}{r'^3} + \frac{3G}{2r'^3} \left[I_{x'y'} + (I_{z'} - I_{x'y'}) \frac{z'^2}{r'^2} \right]$$

or

$$V = - \frac{Gm_1}{r'} - \frac{G}{2r'^3} \Delta I + \frac{3G}{2r'^5} \Delta I z'^2, \qquad (A17)$$

Where $\Delta = I_{z'} - I_{x'y'}$.

The force resulting from this potential is

$$\vec{F} = -\vec{\nabla}'V, \qquad (A18)$$

where $\vec{\nabla}'$ is the del operator in cylindrical coordinates. Hence,

$$\vec{F} = - \frac{Gm_1}{r'^3} \vec{r}' - \frac{3G\Delta I}{2r'^5} \vec{r}' + \frac{15G\Delta I z'^2}{2r'^7} \vec{r}' - \frac{3G\Delta I}{r'^5} \vec{z}'. \quad (A19)$$

Collecting terms,

$$\vec{F} = - \frac{Gm_1 \vec{r}'}{r'^3} - \frac{3G\Delta I}{2r'^5} \left(1 - \frac{5z'^2}{r'^2}\right) \vec{r}' - \frac{3G\Delta I}{r'^5} \vec{z}' \qquad (A20)$$

as stated in (2a).

APPENDIX B

EVALUATION METHOD FOR THE JACOBIAN
DETERMINANTS*

It was mentioned in the paragraph preceding (26) that the Jacobian determinants must be obtained in terms of the orbital elements of m_2. This appendix shows the method employed with an example. Thus, every possible combination of a, e, i, ω, Ω and T, in the Lagrange brackets is made, provided the evaluation is at the time of perihelion passage.

Recall that from (49)

$$p_{12} = a \ (\cos E - e), \qquad (B1)$$

so that $E = 0$ at perihelion.

With $\alpha_1 = a$ and $\alpha_K = e$ for example, the Legendre brackets are $[a, e]$. Then by (B1)

$$\frac{\partial p_{12}}{\partial a} = 1 - e, \quad \frac{\partial p_{12}}{\partial e} = -a, \quad \text{and} \qquad (B2)$$

* See Moulton, op. cit, pp. 388-399.

$$\frac{\partial}{\partial a} \left(\frac{dp_{12}}{dt} \right) = \frac{\partial}{\partial e} \left(\frac{dp_{12}}{dt} \right) = 0. \qquad (B3)$$

And the Jacobian

$$\frac{\partial (p_{12}, \ dp_{12}/dt)}{\partial (a, \ e)} = 0. \qquad (B4)$$

Once again from (49)

$$h_{12} = a \sin E \sqrt{1 - e^2} \qquad (B5)$$

so at perihelion

$$\frac{\partial h_{12}}{\partial a} = \frac{\partial h_{12}}{\partial e} = 0, \ etc. \ . \ . \ . \qquad (B6)$$

Thus

$$\frac{\partial (h_{12}, \ dh_{12}/dt)}{\partial (a, \ e)} = 0. \qquad (B7)$$

The Jacobians of (B4) and (B7) used in the determinant of PART V yield

$$[a, \ e] = 0 \qquad (B8)$$

Following this method of evaluation, the only non-vanishing brackets are

$$[\Omega, a] = \tfrac{1}{2} na \cos i \sqrt{1 - e^2},$$

$$[\omega, a] = \tfrac{1}{2} na \sqrt{1 - e^2},$$

$$[e, \Omega] = \frac{na^2 e \cos i}{\sqrt{1 - e^2}}, \qquad (B9)$$

$$[e, \omega] = \frac{na^2 e}{\sqrt{1 - e^2}},$$

$$[i, \Omega] = na^2 \sin i \sqrt{1 - e^2}, \text{ and}$$

$$[a, T] = \tfrac{1}{2} n^2 a,$$

where $n = \sqrt{G (m_1 + m_2)}/a^{3/2}$.

For the evaluation of the brackets at aphelion the resulting expressions are identical to those given by (B9), except for a negative sign just to the right of the equality symbol of each expression.

APPENDIX C

ORBITAL ELEMENTS OF THE SOLAR FAMILY*

		Mercury			Venus	Earth
a	=	0.387099	a_j	=	0.723332	1.0
e	=	0.205628	e_j	=	0.006787	0.016722
i	=	7.004167	i_j	=	3.394444	0.0
ω	=	28.752600	ω_j	=	54.373056	101.219720
Ω	=	47.145833	Ω_j	=	75.779720	0.0
m_2	=	0.0554	m_j	=	0.8150	1.0

		Mars	Jupiter	Saturn
a_j	=	1.523691	5.202803	9.538840
e_j	=	0.093377	0.04845	0.055650
i_j	=	1.850	1.304722	2.489440
ω_j	=	285.431940	-86.720833	-21.691667
Ω_j	=	48.786389	99.44167	112.78889
m_j	=	0.1075	317.83	95.147

		Uranus	Neptune	Pluto
a_j	=	19.1819	30.0578	39.44
e_j	=	0.04724	0.00858	0.25
i_j	=	0.77306	1.77278	17.16667
ω_j	=	98.055	-84.01444	113.26667
Ω_j	=	73.47833	130.68111	109.73333
m_j	=	14.54	17.23	0.17

*a and a_j are measured in AU, m_2 and m_j in MU and i, i_j, ω, ω_j, Ω, in degrees.

APPENDIX D

This appendix lists the constants used in the two programs which calculated the time rates of change of Mercury's orbital elements from the expressions given by (61). The first program computed the time rates of change when the positions of the planets are exactly at their respective perihelia. The second program performed the same function, but did it so for the aphelion case.

The units employed for the calculations are the astronomical unit (AU), the century (Ce), and the Earth mass unit (MU). Their relationship with the standard (mks) system of units is

$$1AU = 149.6 \times 10^9 \text{ m},$$

$$1Ce = 3.1557 \times 10^9 \text{ sec},$$

$$\text{and} \quad 1MU = 5.976 \times 10^{24} \text{ kg}.$$

Hence, the gravitational constant G becomes equivalent to

$$G = 1.18563 \ \frac{AU^3}{MU \ Ce^2},$$

while   the   mass   and   the   equatorial
radius of the Sun equal

$$m_1 \quad = \quad 3.32831 \times 10^5 \text{ MU,}$$

$$R_e \quad = \quad 4.65234 \times 10^{-3} \text{ AU,}$$

respectively.

BIBLIOGRAPHY

Allen, C. W. *Astrophysical Quantities*. London: Athlone Press, 1973.

Auwers, A. *Astron. Nachr.*, 128 (1891), 367.

Barnes, T. G., Pemper, R. R., and Armstrong, H. L. "A Classical Foundation for Electrodynamics." *Creation Research Society Quarterly*, 14:1 (1977), 38-45.

Dicke, R. H. "Oblateness of the Sun and Relativity." *Science*, April 26 (1974), 419.

Dicke, R. H., and Goldenberg, H. M. "Solar Oblateness and General Relativity." *Physical Review Letters*, 18:9 (1967), 313-316.

Dingle, H. "The Doppler Effect and The Foundations of Physics" (I) and (II). *British Journal for the Philosophy of Science XI*, Vol. 41, 11-31, and Vol. 42, 113-129.

Gilvarry, J. J., and Sturrock, P. A. "Solar Oblateness and the Perihelion Advances of Planets." *Nature*, 216 (1967), 1283-1285.

Goldstein, H.   *Classical Mechanics*.
    New York: Addison-Wesley, Inc.,
    1950.

Kantor, W.   *Relativistic Propagation
    of Light*.   Lawrence, Kansas:
    Coronado Press, 1976.

McCuskey, S. W.   *Introduction to Ce-
    lestial Mechanics*.   New York:
    Addison-Wesley, Inc., 1963.

Moulton, F. R.   *An Introduction to
    Celestial Mechanics*.   New York:
    Longmans and Co., 1914.

Newcomb, S.   *Elements of the Four
    Inner Planets*. Government Print-
    ing Office, Washington, D.C.,
    1895.

Poor, C. L.   "The Figure of the Sun."
    Astrophysical Journal, *22* (1905),
    103-114, 305-317.

Poor, C. L.   *Gravitation vs. Relativ-
    ity*. New York: G. P. Putnam and
    Sons, 1925.

Price, M. P., and Rush, W. F.   "Non-
    relativistic   Contribution   to
    Mercury's Perihelion Precession."
    American Journal of Physics, 47:6
    (1979), 531-534.

Ramirez, Francisco. *Secular Variations on the Orbital Motion of Mercury*. M. S. Thesis, University of Texas at El Paso, August 1982.

Ramsey, A. S. *Dynamics, Part I*. Cambridge, 1943.

Slusher, H. S. "Cosmology and Einstein's Postulate of Relativity." *Creation Research Society Quarterly*, 17:3 (1980), 146-147.

Smart, W. M. *Celestial Mechanics*. New York: Longmans and Co., 1953.

Waldron, R. A. *The Wave and Ballistic Theories of Light-A Critical Review*. Frederick Muller Limited, 1977.